AutoUni – Schriftenreih

Volume 170

Series Editor

Volkswagen Aktiengesellschaft, Volkswagen Group Academy, Volkswagen Aktiengesellschaft, Wolfsburg, Germany

Friedrich von Bülow

A Data-Driven Fleet Service: State of Health Forecasting of Lithium-Ion Batteries

Springer Vieweg

Friedrich von Bülow
Volkswagen AG
Wolfsburg, Germany

Dissertation, angenommen von der Bergischen Universität Wuppertal, Deutschland, 2023
Doctoral thesis, accepted by the University of Wuppertal, Germany, 2023

Ergebnisse, Meinungen und Schlüsse dieser Publikation sind nicht notwendigerweise die der Volkswagen Aktiengesellschaft.
The results, opinions and conclusions expressed in this publication are not necessarily those of Volkswagen Aktiengesellschaft.

ISSN 1867-3635 ISSN 2512-1154 (electronic)
AutoUni – Schriftenreihe
ISBN 978-3-658-43187-7 ISBN 978-3-658-43188-4 (eBook)
https://doi.org/10.1007/978-3-658-43188-4

This Springer Vieweg imprint is published by the registered company Springer Fachmedien Wiesbaden GmbH, part of Springer Nature.
The registered company address is: Abraham-Lincoln-Str. 46, 65189 Wiesbaden, Germany

Paper in this product is recyclable.

Preface

This doctoral thesis was written as part of my work as doctoral student during the three-year doctoral program within Group Innovation at Volkswagen AG in cooperation with the Institute for Technologies and Management of Digital Transformation (TMDT) at the University of Wuppertal (BUW). I would like to express my thanks to everyone who encouraged, challenged, and supported me on the way to this dissertation. Previously published work created in the context of this dissertation is outlined below the acknowledgments.

I would like to express my deepest appreciation to Prof. Dr.-Ing. Tobias Meisen who supervised this dissertation. From the inception through the COVID-19 pandemic and many other challenges, his consistent guidance, constructive feedback, and extensive experience has been exceptional.

Likewise, I am obliged to Prof. Dr.-Ing. Matthias Brockmann from FH Münster for his dedication as 2nd assessor of this dissertation.

This dissertation would not have been possible without the support and nurturing of Dr.-Ing. Oliver Cassebaum. His helpful advice and valuable experience laid an comprehensive foundation for this dissertation. My special thanks go to my colleagues and doctoral students for all the discussions and the fruitful cooperation not only in the context of this dissertation, but especially beyond it. They enable me to realize the full potential of a PhD in a practical industrial context.

In particular, I would like to express my gratitude to Dr.-Ing. Felix Heinrich for the collaboration and multifaceted discussions that enabled a smooth launch phase of this dissertation.

I extend my special thanks to the Master's students Joshua Mentz, Hendrik Schleißheimer, and Markus Wassermann, whom I had the honored to supervise. Our collaboration provided valuable insights that are also reflected in this dissertation.

Furthermore, I would like to thank the Volkswagen PhD Network (Doktorandenkolleg) for the joint companionship through the dissertation, which gave a special sparkle to these three years.

I am also grateful to the research assistants and the team at TMDT for the exchange and critical discussions during colloquia and journal clubs. In particular, I had the great pleasure of working closely, but remotely, with Yannik Hahn.

I wish to sincerely thank the training squad at the Wolfsburg Rowing Club and the trombone band for providing the right balance, both physically and mentally.

Finally, I am extremely grateful to my friends, my family, my parents, and my brother. Through encouragement and profound belief in my abilities, they enabled this thesis long before it really started.

Wolfsburg
March 2023

Friedrich von Bülow

Context of this Thesis

Partial results of this thesis were published in advance in journals and conferences. The following overview shall provide transparency in which chapters of this thesis content from previous publications appears. The publications are also mentioned at the beginning of the respective chapters. Furthermore, the listed patents and disclosures were created related to this thesis.

Journal Publications

Ref.	Publication	Chapters
[1]	Friedrich von Bülow, Joshua Mentz, and Tobias Meisen. "State of health forecasting of Lithium-ion batteries applicable in real-world operational conditions." In: *Journal of Energy Storage* 44 (2021), p. 103439. ISSN: 2352152X. DOI: https://doi.org/10.1016/j.est.2021.103439	2.1, 2.2.1.1, 5.1.1.1, 6
[2]	Friedrich von Bülow and Tobias Meisen. "A review on methods for state of health forecasting of lithium-ion batteries applicable in real-world operational conditions." In: *Journal of Energy Storage* 57 (2023), p. 105978. ISSN: 2352152X. DOI: https://doi.org/10.1016/j.est.2022.105978	2.1, 3, 4,
[3]	Friedrich von Bülow, MarkusWassermann, and Tobias Meisen. "State of health forecasting of Lithium-ion batteries operated in a battery electric vehicle fleet." In: *Journal of Energy Storage* 72 (2023), p. 108271. ISSN: 2352152X. DOI: https://doi.org/10.1016/j.est.2023.108271	2.1, 3.3, 5.2, 8
[4]	Friedrich von Bülow, Yannik Hahn, Richard Meyes, and Tobias Meisen. "Transparent and Interpretable State of Health Forecasting of Lithium-Ion Batteries with Deep Learning and Saliency Maps." In: *International Journal of Energy Research* 2023 (2023), pp. 1–23. ISSN: 1099–114X. https://doi.org/10.1155/2023/9922475.	2.1, 10.4

Conference Publications

Ref.	Publication	Chapters
[5]	Friedrich von Bülow and Tobias Meisen. *State of Health Estimation of Lithium-Ion Batteries under Real-World Operational Conditions. Advanced Battery Power Conference.* 2021. URL: https://2021.battery-power.eu/poster2/512/ (visited on 01.03.2023)	2.1, 3.1
[6]	Friedrich von Bülow, Felix Heinrich, and Tobias Meisen. "Fleet Management Approach for Manufacturers displayed at the Use Case of Battery Electric Vehicles." In: *IEEE International Conference on Systems, Man, and Cybernetics (SMC)* (Melbourne, Australia, 17.10.–20.10.2021). IEEE, 2021, pp. 3218–3225. ISBN: 9781665442084. DOI: https://doi.org/10.1109/SMC52423.2021.965 8680	3.4, 9
[7]*	Friedrich von Bülow, Felix Heinrich, and Tobias Meisen. "A new perspective for manufacturers: Cloud-based utilization of operational fleet data." In: *ELIV 2021* (Bonn, Germany, 20.10.–21.10.2021). Ed. by VDI Wissensforum GmbH. VDI-Berichte 2384. Düsseldorf, Germany: VDI Verlag, 2021, pp. 117–130. ISBN: 978-3-18-092384-0. DOI: https://doi.org/10.51202/9783181023846-117	3.4, 9,
[8]	Friedrich von Bülow and Tobias Meisen. "State of Health Forecasting of Heterogeneous Lithium-ion Battery Types and Operation Enabled by Transfer Learning." In: *PHM Society European Conference* 7 (1 2022), pp. 490–508. DOI: https://doi.org/10.36001/phme.2022.v7i1.3312	2.2.5, 3.4, 4.3, 5.1, 7
[9]†	Friedrich von Bülow, Markus Wassermann, and Tobias Meisen. *State of Health Forecasting of Lithium-Ion Batteries operated in a Mobility on Demand Fleet.* Poster, Batterieforum Deutschland 2023, Berlin, Germany. 2023. URL: https://www.batterieforum-deutschland.de/kon gress/	5.2, 8
[10]†	Friedrich von Bülow. "Lithium-Ion Battery State of Health Forecasting with Machine Learning and Vehicle Fleet Data." In: *Advanced Automotive Batteries Conference Europe (AABC)* (Mainz, Germany, 19.06.–22.06.2023). 2023. URL: https://www.advancedautobat.com/23/aabc-europe/battery-intelligence	5.2, 8

* Short version of [6]. † Poster/Talk corresponding to journal paper [3].

Patents & Disclosures

Ref.	Patents & Disclosures
[11]	Friedrich von Bülow and Felix Heinrich. "Betrieb eines elektrischen Energiespeichers in einem Kraftfahrzeug." G07C 5/08. DE 10 2021 204 847 A1 (Germany). Volkswagen AG. 01.12.2022
[12]	Friedrich von Bülow and Felix Heinrich. "Prädiktion einer Charakteristik einer Zielflotte." G07C 5/08. DE 10 2021 204 849 A1 (Germany). Volkswagen AG. 17.11.2022
[13]	Friedrich von Bülow, Jonas Krog, and Lukas Kille. "Abwärmenutzung einer Schnelllade-Infrastruktur." B60L 53/302. DE 10 2022 201 609 A1 (Germany).Volkswagen AG. 17.08.2023
[14]	Friedrich von Bülow, Fabian Kai-Dietrich Noering, and Daniel Fries. "Verfahren zum Bewerten eines technischen Zustands eines Kraftfahrzeugs mittels eines Bewertungssystems, Computerprogrammprodukt sowie Bewertungssystem." DE 10 2022 213 940.2 (Germany). Volkswagen AG. Publication in process
[15]	Andreas Udo Sass and Friedrich von Bülow. "Verfahren und Vorrichtung zum maschinellen Lernen für die Alterungsvorhersage einer Komponente eines Fahrzeugs." DE 10 2023 201 255.3 (Germany). Volkswagen AG. Publication in process

Abstract

The production capacity for lithium-ion battery (LIB) cells and registrations of new battery electric vehicles (BEVs) are rising worldwide. The range and residual value of BEVs are limited by the state of health (SOH) which is why the SOH and its forecast are in the interest of customers, fleet managers, researchers, and industrial companies. The SOH depends on the LIB's operational load during charging, discharging, and rest phases which can be specified by the time series signals current, temperature, and state of charge (SOC). Deriving features from these three time series signals that encode the operational load for long time windows of aging and the corresponding state of health change (ΔSOH) enables forecasting the battery's SOH. Specifically, a machine learning (ML) model can learn the influence of the operational load on the ΔSOH. Given the limitations of state-of-the-art methods, this thesis presents an SOH forecasting method that is suitable for LIB systems in real-world BEV operation because its histogram-based features can capture the higher operational variability compared to constant laboratory operation. Furthermore, a transfer of an existing model for SOH forecasting to new LIB cell types and new operational domains is conducted which advantageously reduces the necessary amount of new data acquired cutting cost and saving time. The presented SOH forecasting method can be provided as a cloud service via a web or smartphone app to fleet mangers. Forecasting the SOH enables fleet managers of BEV fleets to forecast and plan vehicle replacements. Furthermore, the fleet's operational strategy can be optimized by scenario-based adaption of the features describing the battery operation to reduce the total cost of ownership.

Kurzfassung

Die Produktionskapazität für Lithium-Ion Batteriezellen und auch die Neuzulassungen von rein batterie-elektrischen Fahrzeugen (BEVs) steigen weltweit. Die Reichweite und der Restwert von BEVs sind durch den Gesundheitszustand (SOH) der Batterie limitiert. Deshalb sind der SOH und seine Vorhersage von Interesse für BEV-Nutzer, Flotten-Manager, Forscher und Unternehmen. Der SOH wird von der betrieblichen Belastung der Batterie während des Ladens, Entladens und Ruhezeiten beeinflusst, welche durch die Zeitreihensignale des Stroms, der Temperatur und des Ladezustands (SOC) beschrieben werden können. Die Ableitung von Features aus diesen drei Zeitreihensignalen, welche die betrieblichen Belastung der Batterie über einen langen Alterungszeitraum encodieren, sowie der korrespondierenden Änderung des SOHs ermöglicht die Vorhersage des SOHs einer Batterie. Insbesondere kann ein Machine Learning (ML) Modell den Einfluss der betrieblichen Belastung auf die Änderung des SOH erlernen. Aufgrund der Limitationen bestehender SOH-Vorhersagemethoden präsentiert diese Dissertation eine SOH-Vorhersagemethode, die zur Anwendung mit Batteriesystem im Fahrzeugbetrieb von BEVs geeignet ist. Dies wird durch die Wahl von Histogramm-basierten Features ermöglicht, die die höhere betriebliche Variabilität im Fahrzeugbetrieb verglichen mit dem konstanten Laborbetrieb erfassen können. Weiterhin wird ein Transfer eines bestehenden SOH-Vorhersagemodells zu neuen Batteriezelltypen und neuen betrieblichen Domänen durchgeführt, was vorteilhafterweise die notwendige Menge an neuen Daten reduziert, sodass Kosten reduziert werden und Zeit gespart wird. Die vorgestellte Methode zur Vorhersage des SOHs kann Flottenmanagern als Cloud-Service durch eine Web- oder Smartphone-App zur Verfügung gestellt werden.

Die Vorhersage des SOHs ermöglicht es Flottenmanagern einer BEV-Flotte den Ersatz von Fahrzeugen vorherzusagen und zu planen. Weiterhin kann die Betriebsstrategie der BEV-Flotte optimiert werden, indem die den Batteriebetrieb beschreibenden Features szenariobasiert adaptiert werden, um die Gesamtkosten der Flotten zu reduzieren.

Contents

Acronyms

4G	fourth-generation
5G	fifth-generation
AC	alternating current
ACES	autonomous, connected, electric, and shared
Adam	adaptive moment estimation
ADC	adjusted determination coefficient
AE	absolute error
ANN	artificial neural network
AUC	area under the curve
AULC	area under the learning curve
BaaS	batteries as a service
BASE	Bielefeld Academic Search Engine
BC	boost charging
BEMU	battery electric multiple unit
BESS	battery energy storage system
BET	battery electric truck
BEV	battery electric vehicle
BiSCo	Binary Shift Compression
BJB	battery junction box
BMS	battery management system
BNN	Bayesian neural network
BOL	beginning of life
CAN	controller area network

CC	constant current
CC-CV	constant current–constant voltage
CHP	combined heat and power
CIFAR	Canadian Institute for Advanced Research
CL	Closed-Loop
CNN	convolutional neural network
C_{nom}	nominal capacity
COVID-19	coronavirus disease 2019
CPU	central processing unit
C-rate	current-rate
CTP	cell-to-pack
CV	constant voltage
DC	direct current
DD	Data-Driven
DL	deep learning
DOD	depth of discharge
DOE	design of experiments
DT	decision tree
DV	differential voltage
ECM	equivalent circuit model
ECU	electronic control unit
EFC	equivalent full cycle
EIS	electrochemical impedance spectroscopy
ELU	exponential linear unit
E_{nom}	nominal total energy
EOL	end of life
EOW	end of warranty
E-rate	energy-rate
EU	European Union
EUCAR	European Council for Automotive R&D
EV	electric vehicle
FCR	frequency containment reserve
FEC	full equivalent cycle
FNN	feedforward neural network
FO	few old

FT	Fourier transform
GA	genetic algorithm
GDPR	General Data Protection Regulation
GHG	greenhouse gas
GPR	Gaussian process regression
GPU	graphics processing unit
GRU	gated recurrent unit
GTR	Global Technical Regulation
HD	heavy duty
HEV	hybrid electric vehicle
HGB	Handelsgesetzbuch
HistXGBoost	histogram-based gradient boosting
HPC	high power charging
HV	high voltage
I	identical
IAS	International Accounting Standard
IC	incremental capacity
ICE	internal combustion engine
IEC	International Electrotechnical Commission
IEEE	Institute of Electrical and Electronics Engineers
IoT	internet of things
ISEA	RWTH Aachen ISEA
ISO	International Standards Organization
IUPAC	International Union of Pure and Applied Chemistry
LAAM	loss of active anode material
LACM	loss of active cathode material
LARS	least-angle regression
lasso	last absolute shrinkage and selection operator
LCO	lithium cobalt oxide
LFP	lithium iron phosphate
LIB	lithium-ion battery
LLI	loss of lithium inventory
LMO	lithium manganese oxide
LR	linear regression
LSTM	long short-term memory

LTE	long-term evolution
LTO	lithium titan oxide
LV	low voltage
MaaS	Mobility-as-a-Service
MAD	mean absolute deviation
MAE	mean absolute error
MAPE	mean absolute percentage error
MCC	multistage constant current
MCS	megawatt charging system
MFP	multivariable fractional polynomial
ML	machine learning
MLE	maximum likelihood estimation
MLP	multilayer perceptron
MN	many new
MNIST	Modified National Institute of Standards and Technology database
MOD	mobility on demand
MSE	mean squared error
MTL	multi-task learning
NaN	Not a Number
NARX	nonlinear autoregressive dynamic network with external inputs
NASA random	NASA Randomized Battery Usage Data Set
NCA	lithium nickel cobalt aluminum oxide
NEDC	New European Driving Cycle
NEDO	New Energy and Industrial Technology Development Organization
NEFZ	Neuer Europäischen Fahrzyklus
NI	non-identical
NLP	natural language processing
NMC	lithium nickel manganese cobalt oxide
OCV	open-circuit voltage
ONM	old new mixed
OTA	over the air

Oxford	Oxford Path Dependent Battery Degradation Data Set
p	parallel
PCS	process control system
PGNN	physics-guided neural networks
PHEV	plug-in hybrid electric vehicle
PRD	product requirements document
ps	parallel-series
PV	photovoltaic
R^2	coefficient of determination
RC	relative capacity
RCL	relative capacity loss
RDC	representative driving cycle
ReLU	rectified linear unit
ResCNN	convolutional neural network with residual blocks
ResMLP	multilayer perceptron with residual blocks
RF	random forest
RFR	random forest regression
RMSE	root mean squared error
RNN	recurrent neural network
R_{nom}	nominal resistance
RQ	research question
RSF	random survival forest
RUL	remaining useful life
RV	residual value
s	series
SBM	similarity-based modeling
SCADA-System	supervisory control and data acquisition system
SD	standard deviation
SEI	solid-electrolyte interphase
seq2seq	sequence to sequence
SI	International System of Units
SIB	lithium-sulfur battery
SLSTM	stacked long short-term memory
SMBO	sequential model-based global optimization
SO	source only
SoaP	state of available power

SOB	state of balance
SOC	state of charge
SOE	state of energy
SOF	state of function
SOH	state of health
SOH_{C}	capacity-based state of health
SOH_{E}	energy-based state of health
SOH_{R}	resistance-based state of health
SOP	state of performance
SOT	state of temperature
SotA	state-of-the-art
sp	series-parallel
SSB	solid-state-battery
SSE	sum of squared explanations
SSR	sum of squared residuals
SST	sum of squared totals
STM	source target mixed
SVM	support-vector machine
SVR	support-vector regression
tanh	hyperbolic tangent
TCN	temporal convolutional network
TCO	total cost of ownership
TO	target only
TPE	tree-structured parzen estimator
UBE	usable battery energy
UDDS	Urban Dynamometer Driving Schedule
UN	United Nations
USABC	United States Advanced Battery Consortium
V2C	vehicle to cloud
V2G	vehicle to grid
Val	validation
VC	variable current
vMLP	variational multilayer perceptron
WLTP	Worldwide harmonized Light vehicles Test Procedure
w_{s}	window shift

w_{w}	window width
XGBoost	eXtreme Gradient Boosting
ΔSOC	state of charge lift
ΔSOE	state of energy lift
ΔSOH	state of health change

List of Figures

List of Tables

Introduction

1

1.1 Motivation

Reaching the global reduction of greenhouse gas (GHG) emissions formulated in the Paris Agreement requires adjustments to the transportation sector [16]. The European Commission specifies the GHG reduction through the objective of climate neutrality by 2050 among others by increasing the share of electric vehicles (EVs) [17; 18]. This strategy results in increasing EV sales, not only in Europe, but also in the large automotive markets of the US and China. In fact, already from 2018 to 2022 the worldwide EV stocks increased five-fold in four years to over 26 million EVs. This trend is expected to continue considering the commitment of major automotive manufacturers to their battery supply chain and battery research [19–21; 22, p. 141; 23, p. 2].

In EVs lithium-ion batterys (LIBs) are currently the most important battery technology because of their superiority regarding energy density, power density, lifetime, and cost [22, p. 141; 24, p. 2351; 25, p. 14]. Before EVs, LIBs were first established as popular electrochemical power source in portable applications like consumer electronics starting with their commercialization by Sony Corporation in 1991 [25, p. 14; 24, p. 2351; 26, p. 8; 27, p. 29]. Already by 2000 the majority of notebooks were powered by LIBs [28]. But LIBs are also used in stationary applications, e.g., for stabilization of the grid voltage [24, p. 2351; 26, p. 8].

17 years after the first commercialization of LIBs in 1991, the first EV in series production using LIBs was the Mitsubishi i-MIEV in 2008 [23, p. 9; 29, p. 621]. In the automotive application, LIBs in battery electric vehicles (BEVs) are used in a new context with: 1. More demanding technical requirements, 2. large serial-parallel connections of battery cells [30], and 3. thus, a higher cost share of the battery compared to the overall product than in consumer electronics. That is why

F. von Bülow, *A Data-Driven Fleet Service: State of Health Forecasting of Lithium-Ion Batteries*, AutoUni – Schriftenreihe 170,
https://doi.org/10.1007/978-3-658-43188-4_1

the batteries of BEVs attracts attention by customers, industrials, especially start-ups, and researchers [31; 32].

These more demanding technical requirements for LIBs of the automotive application are [24; 29; 33; 34, p. 24–27; 35, p. 835]:[1] 1. Energy and power, 2. lifetime (calendar life and cycle life), 3. cost, 4. safety (mechanical, thermal, and electrical), and 5. fast charging capability. Especially the high sensitivity of LIBs to operational conditions needs consideration, e.g., during fast charging, ensuring a safe temperature range, management and control of all cells in the high voltage (HV) battery system [24]. Compared to hybrid electric vehicle (HEV) and plug-in hybrid electric vehicle (PHEV), these requirements are even higher for BEVs because the battery is their only power source.[2]

Given the higher energy and power requirements of BEV's, more battery cells are needed compared to consumer applications and to stationary batteries [37, p. 461]. Thus, the battery cost share is also higher: In 2020, the HV battery was the most valuable component of BEVs with 39 % of the total vehicle costs. It is estimated to be still at 20 % to 25 % in 2030 [32; 38], while it accounts for only 1 % to 4 % of the total cost in smartphones [39; 40].

That, is why the residual value (RV) of LIBs is of special interest for private customers, industrials, and fleet managers, e.g., of mobility on demand (MOD) fleets such as carsharing or ride sharing services.[3] BEVs will become even more part of shared mobility concepts, especially in urban areas [43; 44]. As driver-vehicle assignment is dynamic, the operators of such shared BEV fleets are responsible for operation and maintenance of the vehicles. However, the large size of such fleets prevents human operation and maintenance control. The future trend of autonomous cars will also aggravate this problem. Therefore, fleet operators demand services for the health of vehicles and their components, like the battery.

The previously mentioned technical requirements need to be fulfill during the complete calendar life of at least 10 years of BEVs, compared to only 4 years to 7 years for consumer electronics [45].[4] One crucial technical requirement for BEV users and owners is the battery's energy content, i.e., capacity, which determines the vehicle's range [48]. However, when a battery ages the capacity decreases depending

---

[1] For example [33] require for BEV: Calendar life (15 years), cycle life (1000 cycles), temperature range (−30°C to 52°C), and cost on system level ($125/kWh).

[2] For definitions of HEV, PHEV, and BEV consider [36, pp. 6–7].

[3] For a differentiation of mobility on demand (MOD) and Mobility-as-a-Service (MaaS) consult [41; 42].

[4] USABC (US) [33]: 15 years and 1000 cycles; EUCAR (Europe) [46]: BEV life equals 150 000 km and 22 MWh to 24 MWh; NEDO (Japan) [47]: 10 years to 15 years and 1000 cycles to 1500 cycles [29, p. 622].

on its usage and environmental conditions [49–51]. Commonly, the current state of capacity degradation is expressed as state of health (SOH) comparing the current capacity to the capacity of a new battery at its beginning of life (BOL). Hence, the control of the future battery aging caused by the BEV's usage is indispensable.

In the literature, two methodological principals are observed: SOH estimation and SOH forecasting. First, SOH estimation is the determination of the $SOH(t)$ at a certain point in time t. SOH estimation for future SOH is often called SOH prediction in the literature when applicability in the future is doubtful. Second, battery aging is a state change of the SOH from a current $SOH(t_1)$ to a future $SOH(t_2)$ due to aging causes. SOH forecasting is predicting this state change (State of health change (ΔSOH)) or the future $SOH(t_2)$ given information on the battery operational load during the period of time $\Delta t = t_2 - t_1$ until the future point in time t_2. Thus, compared to SOH forecasting, the SOH estimation does not model a change in state, but only a determination of state.

Much research has been conducted to estimate the current SOH (diagnosis), e.g., based on current electrical behavior of the battery [52–57]. Other researchers have designed experiments to examine battery aging, i.e., SOH degradation, under laboratory conditions to identify operational conditions which induce accelerated battery aging [57–59]. However, of great interest for private BEV operators, fleet managers, and automotive manufacturers is the future SOH (prognosis), preferably not after a single cycle or a few days, but after multiple cycles, weeks or months when the battery has aged considerably [60, p. 6]. The future SOH depends on the future battery usage and environmental conditions [61].

SOH forecasting can be of interested for the following stakeholders: Fleet managers, private BEV operators, operators of batteries in 2nd life applications, insurances, and battery designers.

A fleet manager desires to plan replacements of the BEVs in her fleet once batteries do not fulfill their requirements anymore. Also, adapting the scenario of future battery usage enables deriving an operational strategy to control deceleration or acceleration of battery aging. The SOH forecasting model could also be embedded into optimization algorithms for determining the operational strategy. The operational strategy may focus on charging, discharging, or parking. For charging, a charging policy for charging management may be derived. For instance, when discharging, the maximum discharge current can be limited above a certain temperature threshold, or a certain driving mode limiting the power output (e.g., sport, comfort, eco) can be specified. For long parking, a suitable state of charge (SOC) can be specified.

Fleet managers and private BEV operators may be interested in how their battery operation influences the future SOH and, thus, the correlated RV at the end of life (EOL) of the battery's 1th life. The causal connection of SOH and RV is the BEV's range and the fast-charging capability of the BEV when a derating strategy reduces the fast-charging current [62].

Additionally, all involved stakeholders may benefit from SOH forecasting during the transition of the batteries from 1st life to 2nd life applications like battery energy storage system (BESS) for grid energy: During 2nd life screening old cells with similar SOH shall be grouped together to be assembled as new battery packs for 2nd life applications [63; 64]. Determining the current SOH is the task of SOH estimation. However, it will be beneficial if the future SOH of the cells stays similar. This can be ensured by SOH forecasting, especially by probabilistic SOH forecasting.

Furthermore, potential 2nd life applications can be distinguished by their operational conditions and an EOL criterion. Given these operational conditions of the 2nd life, users could back-calculate a minimum required SOH after 1st life that ensures meeting the 2nd life EOL criterion. The same way users can adapt the operational conditions in 1st life to simulate possible scenarios to transition to 2nd life. Such a model can also be interesting for insurances to cover battery manufacturers warranties [65]. Whoever issues warranties for battery life and SOH faces provisions for these warranties because of uncertain liabilities.[5] If SOH forecasting can reduce the uncertainty of these liabilities, a reduction of the corresponding provision can be possible which would decrease the expenses.

Furthermore, battery designers can conduct virtual battery aging experiments by adapting the usage scenario data. After adapting the usage scenario data, the model will output its effect on the SOH. This enables a prescriptive analysis and recommendations for an operational strategy which may include a charging policy for charging management.

1.2 Research Questions

Even though SOH forecasting is beneficial for different stakeholders as motivated in Section 1.1, state-of-the-art (SotA) methods for SOH forecasting of LIBs in automotive applications have the following major limitations:

L1 Less variability in data from laboratory operation compared to real-world vehicle operation

Battery data from laboratory operation like [68–71] usually differ significantly from battery data from real-world vehicle operation of BEVs. Differences are the

[5] Provisions are regulated in IAS 37 [66], but also implemented in national law. For instance, in Germany provisions for warranties are mandatory according to §249 Abs. 1 Satz 1 HGB [67].

duration of hold mode with zero current, the variation of charging and discharging currents as well as the operational range of the SOC (state of charge lift (ΔSOC)). Especially due to regenerative braking in automotive applications charging and discharging mode alternate frequently which is called micro-cycling. Also, high load transients appear. However, SotA SOH forecasting models cannot capture the higher variability of real-world battery operation in BEVs compared to laboratory operation.

L2 Assuming same load during whole battery life
The majority of SotA SOH forecasting models assumes the same load during the whole battery life. Nevertheless, batteries cycled in real-world vehicle operation may have a varying load over the life of the battery, e.g., depending on the operator and the operational region.

L3 Based on data from battery cells, not from battery packs
Furthermore, SotA SOH forecasting models are mostly based on data from battery cells [72, p. 362; 73, p. 28]. But in BEVs battery cells are combined to battery packs which have a more complex aging behavior than a single cell because of the interactions of all cells. These interactions cause the degradation process of the battery pack and of the cells in that pack to depend on each other [74].

L4 Transfer to different battery domain
Lastly, transferability of current SOH forecasting models to new battery types or a new operational domain is limited, but in principle possible [2; 75, p. 59]. One challenge is that manufacturers only have few data on system-level of new vehicles, e.g., from test vehicles and endurance testing.

Advancing SOH forecasting models based on battery operation data from laboratory to real-world vehicle origin is becoming feasible now given two developments:

First, registrations of BEVs are rising as stated before. Especially, suitable real-world data for SOH forecasting becomes available on a large scale due to intensive usages of BEVs in fleets for MOD.

Second, the data from these vehicles can now be accessed on a big scale. Sensors exist in vehicles at least since 1975 [76]. These sensors broadcast messages by the electronic control units (ECUs) over bus-based networks, in most cases the controller area network (CAN) which was introduced by Bosch in 1991 [77; 78]. Starting from 2014 first vehicles with fourth-generation (4G) broadband cellular network technology, e.g., long-term evolution (LTE) were connected to the cloud [79]. The exchange of sensor data from the vehicle to cloud (V2C) will further be boosted by the introduction of fifth-generation (5G) networks starting in the 2020s [80–82]. This trend can be seen as part of the rising connectivity of machines in the context of the internet of things (IoT) and Industry 4.0 [83–85]. In the cloud, the transmitted vehicle data can be analyzed in detail and used as base for machine learning (ML) models to provide data-driven services. One such data-driven fleet service is SOH forecasting which faces the challenge that automotive data from the CAN bus typically requires data cleansing and preprocessing because the resulting time series data is asynchronous, irregular, and multi-cardinal [78].

Summarized, the trend of future mobility to autonomous, connected, electric, and shared (ACES) vehicles [86; 87] makes SOH forecasting now more feasible and relevant.

Thus, given the changing data-availability from BEVs and the presented limitations of SotA methods for SOH forecasting four research questions (RQs) are derived:

RQ.1 How can the SOH of LIB cells in laboratory operation be forecasted? (Cell & Lab prognosis)

RQ.2 How can a model for SOH forecasting be transferred to LIBs with a different cell chemistry or different operational load? (Transfer)

RQ.3 In what way is a method for SOH forecasting of LIB cells in laboratory operation suitable to forecast the SOH of a LIB system in vehicle operation? (System & BEV prognosis)

RQ.4 How can SOH forecasting of LIBs in automotive applications technically be realized? (Open RQ)

The encoding of higher operational variability (L1) and different load during the battery life (L2) are first realized for cell data from laboratory operation in RQ.1. Based on this, in RQ.2 the model transfer is examined (L4). RQ.3 progresses from RQ.1 by using data from battery packs (L3) and still considering L1 and L2. As all four limitations are motivated from a practical perspective of BEV fleets, RQ.4 openly discusses possibilities for the technical implementation of an SOH forecasting method as a data-driven fleet service.

1.3 Structure of this Thesis

In the following, the structure of this thesis and the topics dealt within the respective chapters are described. The remainder of this thesis is structured as depicted in Figure 1.1.

Chapter 2 describes the foundations of battery technology and supervised ML models. Chapter 3 differentiates SOH estimation and forecasting as well as remaining useful life (RUL) prediction. Furthermore, scenarios and the model transfer in

SOH forecasting are discussed. The defined model key criteria lead to Chapter 4 that provides a structured literature survey of SOH forecasting methods. Chapter 5 describes the used data sets, both from laboratory and vehicle fleet operation. In the next three Chapters 6 to 8, SOH forecasting is examined in different settings with the corresponding results presented and discussed within each respective chapter: In Chapter 6, SOH forecasting is first implemented using a public LIB cell data set from laboratory operation (RQ.1). In Chapter 7, a model obtained in Chapter 6 is used as

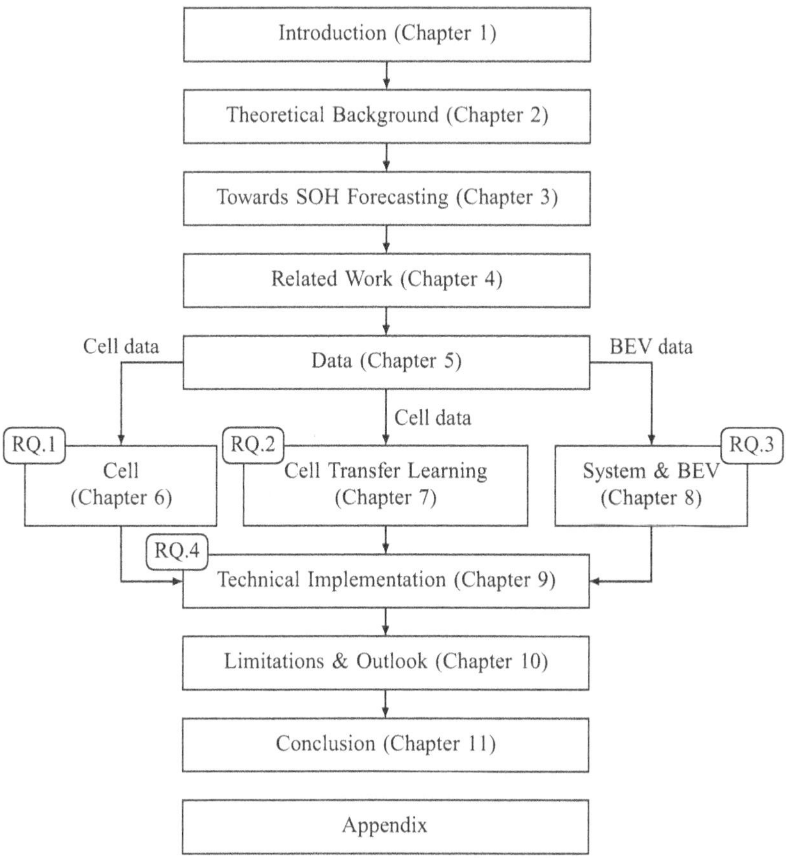

Figure 1.1 Overview of the structure of this thesis and the research questions that are answered within the respective chapters.

base for a model transfer to other laboratory LIB cell data sets (RQ.2). In Chapter 8, the SOH forecasting method from Chapter 6 is applied to a real-world LIB pack data set from MOD BEV fleet operation (RQ.3). Chapter 9 throws light on a conceptual technical implementation for SOH forecasting in BEV fleets (RQ.4). Chapter 10 provides an outlook onto possible future work building upon the limitations of this thesis which is concluded by Chapter 11.

Theoretical Background

2

To answer the RQs, considering the current limitations of SOH forecasting methods introduced in Section 1.2, background from two fields is required as depicted in Figure 2.1: Given the importance of LIBs as power source of BEVs, Section 2.1 provides domain knowledge on LIB cells, their aging, and their composition to battery modules and battery packs. In Section 2.2 relevant basics of supervised ML models for regression are presented. The importance of these two fields is also underlined by the existing data-driven SOH forecasting methods presented in Chapter 4. The background on LIB presented in Section 2.1 is based on previous publications [1–3; 8]. The background on ML presented in Section 2.2 is based on previous work [1; 3; 8].

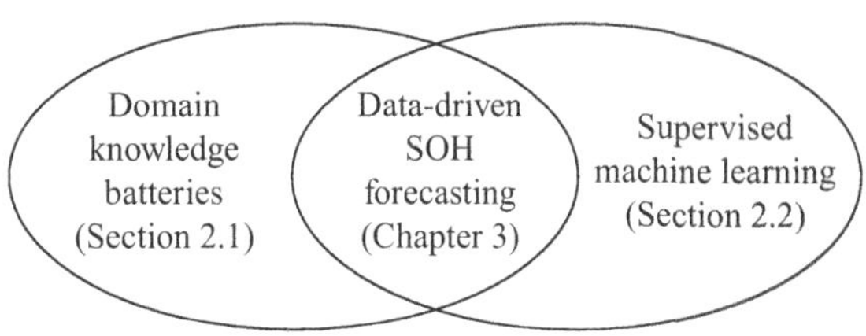

Figure 2.1 Structure of Chapters 2 and 3

Supplementary Information The online version contains supplementary material available at https://doi.org/10.1007/978-3-658-43188-4_2.

F. von Bülow, *A Data-Driven Fleet Service: State of Health Forecasting of Lithium-Ion Batteries*, AutoUni – Schriftenreihe 170,
https://doi.org/10.1007/978-3-658-43188-4_2

2.1 Lithium-Ion Batteries

Batteries are electrochemical energy storage devices that convert chemical energy into electrical energy (discharge). Primary batteries are designed to be discharged only once and then be discarded, i.e., they are non-chargeable. Secondary batteries are rechargeable.[1] For charging, the described discharging process is reversed [88]. A very popular type of secondary battery is the LIB which is used in mobile devices like mobile phones, laptops, and machine tools. In the mobility sector, electric bicycles, electric trains, i.e., battery electric multiple units (BEMUs), and EVs are important applications [25; 89]. Types of EVs are HEVs, PHEVs, and BEVs. For the scope of this thesis, BEVs are the most relevant type of EVs due to the given data in Section 5.2 and the trends mentioned in Chapter 1. For definitions of EV, HEV, PHEV, and BEV consider [36, p. 6].

2.1.1 Operating Principle and Components of Lithium-Ion Battery Cells

Figure 2.2 shows the schematic illustration of a typical LIB cell. The major components are a negative electrode (anode), a positive electrode (cathode), the ion-conducting electrolyte, and the electrically insulating separator. Both electrodes are based on a current collector. Onto each current collector an electro-reactive material mixed with electrically conducting additives is coated. Mobile lithium-ions can be intercalated into the electro-reactive material of the electrode or be extracted from the electrode. The electrolyte is a conducting salt dissociated in an organic, non-aqueous solvent that immerses the electrodes and the separator. The separator is a porous membrane that allows lithium-ions to permeate, but electronically isolates the two electrodes to prevent electrical short circuits [25; 90; 91].

[1] In German, primary batteries and secondary batteries are also distinguished as „Batterie“ and „Akkumulator, Akku“ respectively.

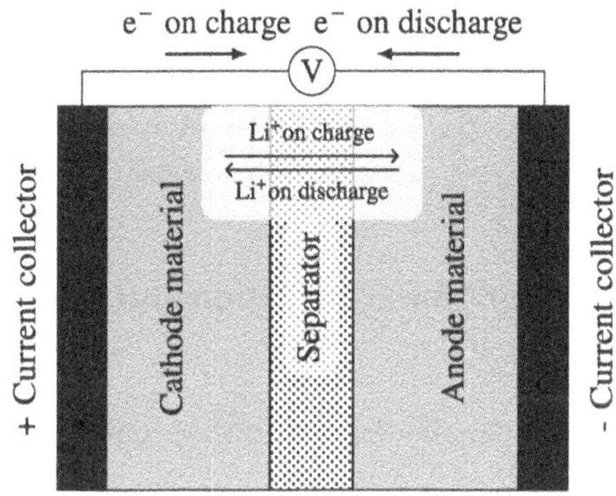

Figure 2.2 Schematic illustration of LIB cell ([92, p. 26.5])

When discharging, lithium-ions diffuse from the anode through the electrolyte and the separator to the cathode due to concentration gradients. Simultaneously, electrons flow from the anode to the cathode through an external electric circuit resulting in an electric current. The oxidation reaction at the carbon anode C_6 is described as:

$$Li_{1+n}C_6 \underset{\text{charge}}{\overset{\text{discharge}}{\rightleftarrows}} n \cdot Li^+ + n \cdot e^- + LiC_6 \tag{2.1.1}$$

with Lithium intercalated in the carbon anode Li_nC_6 oxidizing to n lithium-ions Li^+ and n electrons e^-. The active anode host material LiC_6 remains.

The reduction reaction at an exemplary metal oxide MO_2 cathode is described as:

$$LiMO_2 + n \cdot Li^+ + n \cdot e^- \underset{\text{charge}}{\overset{\text{discharge}}{\rightleftarrows}} Li_{1+n}MO_2. \tag{2.1.2}$$

Lithium-ions Li^+ get reduced by electrons e^- to intercalated lithium on the cathode/electrolyte interface Li_nMO_2.

When charging this process is reversed. The oxidation in Equation (2.1.1) and reduction in Equation (2.1.2) result in the following redox reaction [90, pp. 11–13; 92, p. 26.4]:

$$Li_{1+n}C_6 + LiMO_2 \underset{\text{charge}}{\overset{\text{discharge}}{\rightleftarrows}} LiC_6 + Li_{1+n}MO_2. \tag{2.1.3}$$

The characteristics of a LIB cell like power density and energy, but also the aging behavior depend on the used materials and the form of the cell. Both are briefly introduced in Appendix A.1 (accessible in the Electronic Supplementary Material).

2.1.2 Parameters for Characterization and Operation of Batteries

Different parameters can describe the state and operation of batteries. They are all based on the voltage U in V, current I in A and temperature T in °C. Cell manufacturers usually specify a voltage range with an upper and lower voltage limit, synonymously cut-off voltage, to prevent damaging processes within the battery cell. The same applies for a temperature range [93].

The battery **electrical current**, short current, I is defined in accordance to [94] and the sign convention of International Union of Pure and Applied Chemistry (IUPAC) [95, p. 56] is applied in this thesis with $I_{\text{ch}} > 0\,\text{A}$ for charging and $I_{\text{disch}} < 0\,\text{A}$ for discharging. The current is the change of electrical charge q in a system like a LIB over time [93, p. 412]:

$$I = \frac{\mathrm{d}q}{\mathrm{d}t}. \tag{2.1.4}$$

The amount of electrical charge q a fully charged battery can deliver under specific conditions is called **capacity** C in Ah.[2] For comparability capacity determination requires the same standard operational conditions regarding the temperature, cut-off voltage and charge-discharge profile [97]. Practically, the capacity is the integral of the discharge current $I(\tau)$ over the discharge time τ from a fully charged state until the specified lower cut-off voltage V_{cutoff} at $t_{\text{disch,end}}$ is reached. Then the battery is considered fully discharged. This method is also called coulomb counting [93, p. 413]:

$$C(t) = \int_0^{t_{\text{disch,end}}} I(\tau)\mathrm{d}\tau. \tag{2.1.5}$$

When a battery ages, the capacity decreases and the internal resistance increases. Thus, t refers to the point in battery life when the capacity $C(t)$ was determined [98, p. 395; 96, 482-01-27; 25; 94].

[2] The International System of Units (SI) unit of capacity is Coulomb C with $1\text{C} = 1\,\text{As}$, but usually ampere hours are used ($1\,\text{Ah} = 3600\,\text{C}$) [96, p. 482-01-27; 94].

A special $C(t)$ is the **nominal capacity (C_{nom})** specified by the battery manufacturer. Usually, it is similar, but due to deviations in the manufacturing process not necessarily the same as the initial capacity $C(t_0)$ at the BOL of a new battery, i.e., after production [99]:

$$C_{nom} \cong C(t_0). \tag{2.1.6}$$

This effect is also called intrinsic cell-to-cell variability that may increase while aging [100]. Sometimes C_{nom} is also referred to as rated or installed capacity [101, 3.4; 94, p. 98].

Commonly, discharge and charge currents can be expressed as **current-rate (C-rate)** which is a multiple or fraction of C_{nom} [101, 3.4; 94, pp. 92, 98]:

$$\text{C-rate} = \frac{I}{C_{nom}}, \tag{2.1.7}$$

In the battery industry and research the short notation xC is accepted, e.g., 5C for a C-rate of 5 [102, p. 29].[3] Discharging a battery with $C_{nom} = 2$ Ah and a C-rate of 1C means that the battery is discharged with a constant current (CC) of 2 A. Discharging the same battery with a C-rate of 0.5C means a current of 1 A. Discharging at a C-rate of 2C means a current of 4 A. The C-rate facilitates easy comparison of currents of battery with different C_{nom} [101, 3.4; 102, pp. 28-29; 94, p. 92; 96, 482-01-55 & 482-03-34; 103; 104, p. 445].

Corresponding, the energy-rate (E-rate) describes the discharge power. An E-rate of 1 E is the discharge power to discharge the entire battery in 1h [105; 101, 3.5].

The relative stored electric charge $q(t)$ at a given time t is defined as **state of charge (SOC)** as visualized in Figure 2.3:

$$SOC(t) = \frac{q(t)}{C(t)} \cdot 100\% \quad 0\% \leq SOC(t) \leq 100\% \tag{2.1.8}$$

with the denominator as the current capacity $C(t)$ and not C_{nom}.[4] Thus, during the whole battery life a fully charged battery has $SOC = 100\%$, while a fully discharged battery has $SOC = 0\%$ [94, p. 88; 98, p. 396; 106; 51, p. 793].

[3] The short notation xC should not be confused with the capacity C and electric charge q noted as x C with the SI unit Coulomb [103].

[4] E.g., [98, p. 396] specify C_{nom} as denominator which is not common.

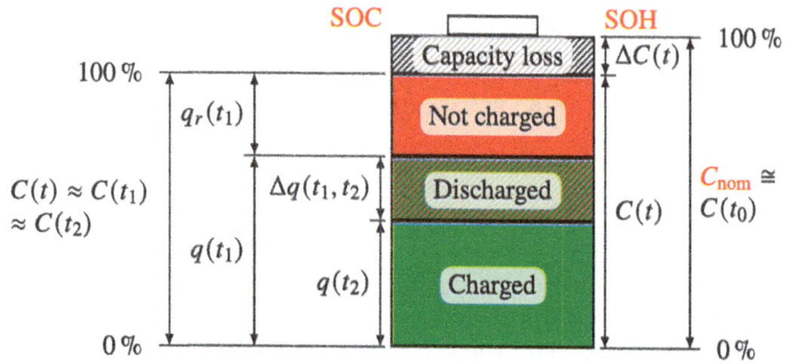

Figure 2.3 Visualization of SOC, ΔSOC, and SOH

Alternatively to the charge-based definition in Equation (2.1.8) with $[C] = \text{Ah}$, the **state of energy (SOE)** defines an energy-based point of view:

$$SOE(t) = \frac{E(t)}{E_{\max}(t)} \cdot 100\% \quad 0\% \leq SOE(t) \leq 100\% \tag{2.1.9}$$

with $E(t)$ as currently stored energy with $[E] = \text{Wh}$ and $E_{\max}(t)$ as maximum possible energy storage. SOE is related to SOC via [94, p. 91]:

$$\begin{aligned} SOE(SOC,t) &= \frac{E(SOC)}{E_{\max}(t)} \cdot 100\% \quad 0\% \leq SOE(SOC,t) \leq 100\% \\ &= \frac{\int_{q(SOC=0\%)}^{q(SOC)} U_{\text{OCV}}(q)\mathrm{d}q}{\int_{q(SOC=0\%)}^{q(SOC=100\%)} U_{\text{OCV}}(q)\mathrm{d}q} \cdot 100\% \end{aligned} \tag{2.1.10}$$

with $U_{\text{OCV}}(q)$ as open-circuit voltage (OCV) of the battery. Full SOE means $SOC = 100\%$ and $SOE = 100\%$. It is the reference value for $E(SOC,t) = E_{\max}(t)$. An empty battery means $SOC = 0\%$ and $SOE = 0\%$, resulting in $E(SOC,t) = 0\,\text{kWh}$ [94, pp. 91–92; 93, p. 420; 107, p. 42].

While the SOC is relative to a full battery, the **depth of discharge (DOD)** is relative to an empty battery. The DOD is the level of the removed electric charge $q_{\text{r}}(t)$ at a given time t [102, p. 64; 51, p. 795; 108, p. 446; 109, p. 24; 110, p. 295]:

$$DOD(t) = \frac{q_{\text{r}}(t)}{C(t)} = \frac{C(t) - q(t)}{C(t)} = 1 - SOC(t). \tag{2.1.11}$$

Both SOC and DOD describe states. The change of SOC during a charging or discharging period from its beginning t_1 to its end t_2 is called **state of charge lift ($\boldsymbol{\Delta SOC}$)** as visualized in Figure 2.3:

$$\Delta SOC(t_1, t_2) = SOC(t_2) - SOC(t_1) = \frac{\Delta q(t_1, t_2)}{C(t)} = \frac{q(t_2) - q(t_1)}{C(t)} \qquad (2.1.12)$$
$$-100\% \leq \Delta SOC \leq +100\%.$$

Evidently, ΔSOC and DOD are not the same: When charging a battery from $SOC(t_1) = 30\%$ to $SOC(t_2) = 60\%$ Equation (2.1.12) leads to $\Delta SOC = 60\% - 30\% = +30\%$. However, the charging process does not have a single DOD of $+30\%$. The charging process can be described by $DOD(t_1) = 100\% - SOC(t_1) = 70\%$ at the beginning and $DOD(t_2) = 100\% - SOC(t_2) = 30\%$ at the end of the charging process.

2.1.3 State of Health and Remaining Useful Lifetime

Due to battery aging characteristic states of the battery change. The **state of health (SOH)** compares the current state of the battery to the state of a new battery at its BOL [49–51]. The SOH can be defined differently depending on the view point, e.g., of a manufacturer or battery user [51]. Any state variable that changes with aging may be suitable for the definition of the SOH, like the number of charging/discharging cycles [111], the voltage change caused by the load of a power or current profile [108], alternating current (AC) impedance, self-discharge rate, and power density [30; 49; 112]. These can be used individually or jointly [112], e.g., as a vector, arithmetic mean, harmonic mean, or geometric mean.

Commonly, the SOH can be described by the increasing internal resistance R, the decreasing capacity C, and the decreasing storable energy E_{max} [113–115]. The **resistance-based state of health ($\boldsymbol{SOH_R}$)** is the current internal ohmic resistance $R(t)$ relative to the nominal internal ohmic resistance R_{nom} [109; 116; 117; 114, p. 3]:

$$SOH_R(t) = \frac{R(t)}{R_{nom}} \cdot 100\% \qquad 100\% \lesssim SOH_R(t). \qquad (2.1.13)$$

Analogous, the **capacity-based state of health ($\boldsymbol{SOH_C}$)** denotes the remaining capacity $C(t)$ relative to the nominal capacity C_{nom} as visualized in Figure 2.3 [114, p. 3; 106, p. 161; 51, p. 795; 93, p. 429; 99; 109]:

$$SOH_C(t) = \frac{C(t)}{C_{nom}} \cdot 100\% \qquad 0\% \leq SOH_C(t) \lesssim 100\% \tag{2.1.14}$$

Further definitions of $SOH_R(t)$ and $SOH_C(t)$ are summarized in Appendices A.2.1.1 and A.2.2.1 respectively (accessible in the Electronic Supplementary Material).

The **energy-based state of health (SOH_E)**[5] denotes the remaining total energy $E_{max}(t)$ of the nominal total energy (E_{nom}) [118, p. 452; 93, p. 429]:

$$SOH_E(t) = \frac{E_{max}(t)}{E_{nom}} \cdot 100\% \qquad 0\% \leq SOH_E(t) \lesssim 100\%. \tag{2.1.15}$$

As [121] state, SOH_C and SOH_E are related via the voltage U because it holds that the energy of a battery is the product of the average discharge voltage $\overline{U_{disch}}$ and the capacity $C(t)$ [25, p. 16]:

$$E_{max}(t) = \overline{U_{disch}} \cdot C(t) \tag{2.1.16}$$

Both, the SOH_C and the related SOH_E reflect the relevant range of BEVs. The SOH_R would be interesting for HEVs and electric trains, i.e., BEMUs, because it reflects the power capability [89; 117; 122–124]. For EV, especially for BEV, the power capability reflects the capabilities for acceleration, regenerative braking, and the gradeability. However, during battery degradation energy capacity fade limits the lifetime more than power fade [125].

The SOH_C and SOC neglect the energy losses of electrochemical reactions and internal resistances inside the battery because they only focus on the charge flow. However, the energy losses may affect the available energy provided to the drive train of a BEV. Thus, the SOH_E and SOE are more meaningful and practically relevant for the operation of BEVs [107; 126]. This is also manifested by BEV manufacturers' common specification of the battery energy content in Wh and not the capacity Ah. Likewise, it corresponds to the legal requirement to specify the energy consumption in kWh/100 km for BEV advertisements in Germany [127]. Compared to the SOE, the SOC is useful for batteries with flat and short discharge voltage transients like lead-acid and alkaline nickel batteries. But the steeper and longer discharge voltage transients of LIBs are only considered in the SOE, not the SOC, e.g., by the average discharge voltage $\overline{U_{disch}}$ in Equation (2.1.16) [93; 128].

[5] Alternatively, Grolleau et al. [118], Guenther et al. [119], and Mamadou et al. [120] use the term SOE. But to avoid confusion with the „energy-based" SOC the notation of SOH_E is used.

Moreover, the new United Nations (UN) Global Technical Regulation (GTR) No. 22 will accelerate importance of the usable battery energy (UBE) and not the capacity as a measure of state because of the importance of the UBE for users, vendors, and buyers of BEVs [129; 130]. Usable energy here refers to the available net energy considering possible limitations of the usable SOC range by the battery management system (BMS) in contrast to the total or gross energy [131; 132]. In the following, the SOH_C is briefly noted as SOH, because it is used in most research work.

The difference and dependence of SOC and SOH are visualized in Figure 2.3. The SOH of a new cell may be below 100% because of the variability of manufacturing processes [133] and the dependence on the specification of C_{nom} in the data sheet of the cell [110, p. 294].

Often SOH and SOC can compensate each other. Hence, none of both alone can describe if the battery currently meets the requirements of a given application in a certain operational situation. Thus, the **state of function (SOF)** is introduced to reflect the capability of the battery performance depending on SOH, SOC, and temperature [134, p. 86; 51, p. 796; 110, p. 295; 50, pp. 166–167; 111].

Other states like the state of available power (SoaP), state of performance (SOP), state of temperature (SOT), and state of balance (SOB) are introduced in [135; 136; 137; 114], but are not relevant for this thesis because they are not necessarily reflecting long-term battery aging. Thus, they are of minor interest in this thesis.

A **cycle** of a secondary battery like a LIB includes discharging and charging. It may also include rest periods with $I = 0\,\mathrm{A}$ which may also be called storage or in the case of an BEV parking without parallel external charging. There exist full and partial cycles: A full cycle has a ΔSOC of 100% discharging from a full to an empty battery and charging back to a full battery, or vice versa. Incomplete discharging and/or charging is called partial cycling. For partial cycling, the amount of partial cycles can be converted to equivalent full cycles (EFCs).[6] E.g., in [140] the conversion is based on the charge throughput in Ah. Therefore the charge throughput is divided by C_{nom}. On the other hand [93, p. 429] use the discharge current instead of the charge current. Both provide better comparability among batteries with different ΔSOC [91, p. 19; 96, 482-03-06].

The **battery life** of a LIB is the period of time between:

1. Beginning of life (BOL): Time of delivery and
2. End of life (EOL): When properties of the LIB fall below an application-dependent threshold due to aging [141, p. 386].

---

[6] The use of the term full equivalent cycles (FECs) is often synonymous as in [138; 139; 37].

Usually in between BOL and EOL the warranty granted by the automotive or cell manufacturer[4] ends determining the end of warranty (EOW). EOL may also correspond to EOW [133].

After EOL of the 1th life, the LIB may be used in a different 2nd life application. These are often applications with less demanding performance requirements like BESS and low requirements regarding the structural assembly space of the battery [37, p. 461]. Before the 2nd life, batteries usually are classified into subgroups of similar performance [64; 49; 142; 143].

Apart from the SOH, that is describing the current battery state referring to the initial, new state at the BOL, there exists another measure of battery aging: The **remaining useful life (RUL)** is referencing to the EOL, not the BOL. Therefore, SOH and RUL are related by their different reference point in the battery life as depicted in Figure 2.4.[7] According to Nuhic et al. [145, p. 681] the RUL is defined as the number of charge-discharge cycles until a specific battery SOH is reached as specified in Equation (2.1.17) with k as number of cycles:[8]

$$RUL_{EOL=80}(t) = k(t_{EOL=80}) - k(t) \tag{2.1.17}$$

and with the common threshold mentioned in the literature for automotive applications at EOL of $SOH_C = 80\ \%$ depicted as red line in Figure 2.4. Below this value, the battery is not considered to be usable for automotive applications anymore [99, p. 117; 106, p. 161; 141, p. 386]. Less common are 65.50 % [147]. However, 80 % as EOL criterion is also questioned because of the increasing maximum capacity and, as a consequence, the increasing range of BEVs [142, p. 2; 143, p. 2711].

Another EOL criterion could be the total lifetime of the battery [148]. Also, EOL may be defined by a SOH_R of 200 % [117; 149] or 160 % [122] when following the definition in Equation (2.1.13). Also, the knee point is suggested as EOL criterion [150]. Hitting the EOL criterion does not imply that the battery cannot be used anymore, it only indicates that the predefined threshold has been surpassed [115]. In practice, manufacturers and users may define their own EOL criteria depending on their battery application [125; 151]. For example, different BEV users may require different ranges leading to different EOL criteria. Alternatively, they could relate to the general concept of failure assessed by criteria like unsatisfying performance, incapability of providing certain functionality, and temporal availability [152; 153].

---

[7] Plotting the RUL over the cycle number or EFC like in [144, p. 7] is very rare.

[8] An alternative equation includes the current cycle in the RUL, i.e., RUL before the current cycle: $RUL_{EOL=80}(t) = k(t_{EOL=80}) - k(t) + 1$ [146].

The RUL usually refers to EFC, but real-world operational battery cycles have different ΔSOC as mentioned in Table 2.3. In the operation of BEVs, the RUL can therefore only serve as an indicator for the number of potentially non-full cycles until the EOL.

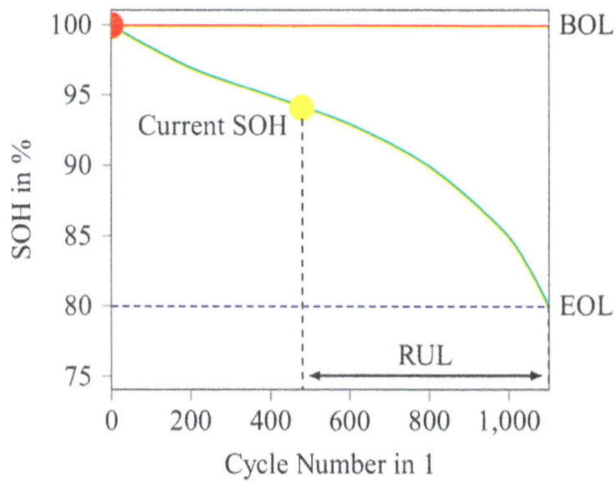

Figure 2.4 Relation of SOH and RUL with their different reference points to BOL and EOL respectively

2.1.4 Fundamentals of Lithium-ion Battery Cell Aging

As introduced in Equations (2.1.13) and (2.1.14) the SOH can refer to decreasing cell capacity (capacity fade) and increasing impedance (power fade) respectively [90, p. 21; 99]. Capacity and power face are the effects of degradation as shown in Figure 2.5.

The effects of degradation base on the three commonly known degradation modes:

1. **Loss of lithium inventory (LLI)** means that the amount of lithium-ions available for diffusion between the cathode and anode reduces. This loss is caused by parasitic reactions, like lithium plating, surface film formations, and decomposition reactions, that consume lithium-ions during their formation and growth.
2. **Loss of active anode material (LAAM)**: As a result of particle cracking and loss of electrical contact, active mass of the anode is no longer available for lithium (de-)intercalation [154].

3. **Loss of active cathode material (LACM)**: Like for LAAM at the anode, the active mass of the cathode available for lithium (de-)intercalation decreases. Causes are structural disordering, particle cracking, and loss of electrical contact.

These three degradation modes are linked to several degradation mechanisms as shown in Figure 2.5. The most common degradation mechanisms are [154-156]:

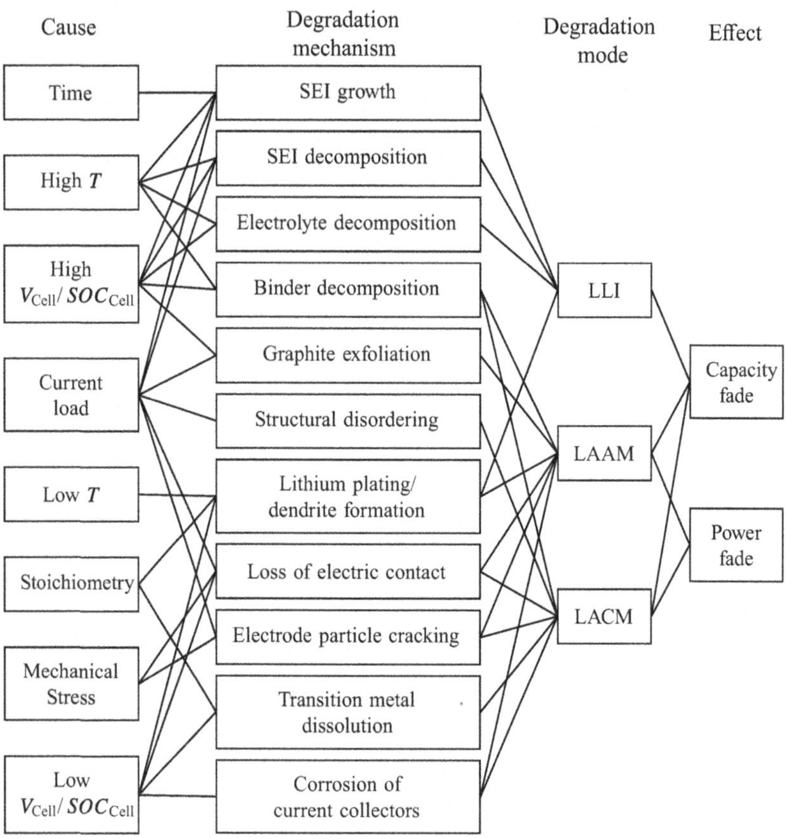

Figure 2.5 Causes, mechanisms, and effects of battery aging [154, p. 17]

1. Solid-electrolyte interphase (SEI) growth: The SEI is a surface film on the anode that forms already at BOL. This interface is attributed to the electrochemical reduction of the solvent and electrolyte salt, generating a mixture of lithium oxide, lithium fluoride, and semi-carbonates between anode and electrolyte.
2. Lithium corrosion: In the active carbon of the anode lithium may corrode causing an irreversible loss of mobile lithium-ions. This leads to capacity fade.
3. Lithium plating: This reaction reduces lithium-ions dissolved in the electrolyte to metal lithium at the surface of the anode's active material while charging. This reaction is partially reversible because plated lithium dissolves to intercalate into the anode active material during rest periods or discharging, but an irreversible part can remain. It advances at high currents and low temperatures while charging [91; 157]. However, Waldmann et al. [158] observed that only combinations of critical temperatures, SOC, and currents promote lithium plating.
4. Loss of electric contact: The SEI becomes detached from the anode resulting in a loss of electrical contact and increasing impedance.

The degradation mechanisms from Figure 2.5 are linked to the battery components anode, cathode, and separator as categorized in Table 2.1. Further, they can be linked to stressors, also called stress factors. Stressors are combinations of operation conditions that induce or accelerate battery aging. These operation conditions are specified by temperature, current, SOC, etc. [109, p. 91]. Known combinations with a high effect on aging are for example overcharging and cycling at partial SOC [61]. Even though these battery stressors are qualitatively known as displayed in Table 2.1, their single and joint impact on the battery's SOH has not been modeled quantitatively in automotive battery systems operated in BEVs yet.

From an operational point of view battery aging can be classified into two categories: Calendar aging and cyclic aging. Calendar aging is associated to the storage of batteries, meaning no charging or discharging is applied. Hence, it is also called passive aging. Cyclic aging is associated to cycling of the battery, i.e., to charging and discharging [58; 57, p. 682]. High temperatures and high SOCs are causing fast battery calendar and cyclic aging [160; 161]. For example, a high SOC over 80 % accelerates SEI growth [57]. In addition, new SEI formation appears after high changes of an electrode's volume due to a significantly low or high SOC [154; 156; 162]. Further, subzero temperatures accelerate lithium plating and capacity fade [161]. Other stressors accelerating battery aging are high charge and discharge C-rates as well as a high ΔSOC [58; 163]. Recovery effects from aging are also possible due to long rest periods during high current cycling [164–166].

Apart from the stress factors, i.e., the operational states of the battery, also their order can influence battery aging. This is known as path dependence [167; 168], especially for cyclic aging with high C-rates [169] whereas path independence of SOC and T was observed for calendar aging [170; 171].

For this thesis, three characteristic battery aging behaviors are distinguished by the form of their SOH trajectory: Linear, superlinear, and sublinear degradation as schematically illustrated in Figure 2.6. Linear degradation is characterized by a constant aging rate over the whole battery life (e.g., NASA Randomized Battery Usage Data Set (NASA random) in Figure 5.1d). In contrast, superlinear degradation is slowly first, but accelerates. Vice versa is sublinear degradation which is often associated with SEI growth, which is self-passivating and, thus, self-limiting [172; 173].

Table 2.1 Degradation mechanisms and their accelerating stressors [154; 159, p. 12; 1]

Battery component	Aging Mechanisms	Accelerated by
Anode	SEI formation	↑ & ↓ SOC
	SEI growth	↑ T, ↑ SOC
	SEI decomposition	↑ C-rate, ↓ T
	Electrolyte decomposition	↑ T, ↑ SOC
	Structural disordering	↑ C-rate, ↑ & ↓ SOC
	Lithium plating	↑ C-rate, ↓ T, ↑ SOC
	Corrosion and loss of electrical contact	↓ SOC
Separator	Blocked pores (Separator and Electrodes)	↑ T, ↑ SOC
Cathode	Binder decomposition	↑ T, ↑ SOC
	Structural disordering	↑ C-rate, ↑ T
	Dissolving of transition metals	↑ T
	Corrosion and loss of electrical contact	↑ C-rate, ↑ T, ↑ SOC

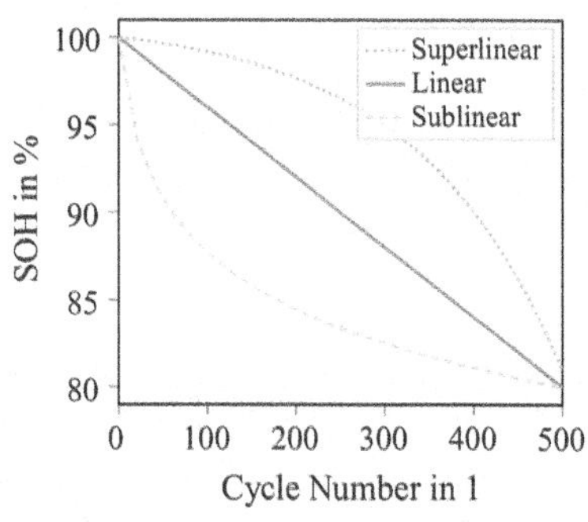

Figure 2.6 Schematic types of battery aging trajectories (based on [173])

Warnecke [174] specifies even more detailed three aging phases of lithium nickel manganese cobalt oxide (NMC) LIBs which are shown in Figure 2.7a: First, a slight increase of SOH is observed. Second, a phase of linear degradation is identified which is dominated by SEI growth [175]. Third, the degradation behavior changes from linear to nonlinear, ending in a rapid break down. This transition is called aging knee or knee point. The transition is caused by new degradation mechanisms, especially lithium plating, becoming more dominant [175; 176]. Fermín-Cueto et al. [172] propose the knee onset as beginning of the accelerated degradation phase, i.e., the end of the linear degradation. The identification of the aging knee is seen as uncritical for EVs because the batteries reach 50 % SOH earlier. This makes the aging knee especially relevant for 2nd life applications [176]. Based on the three aging phases of Warnecke [174], Lucu et al. [285] also mention a 4th phase with a slowdown of capacity loss described by [178] visible in Figure 2.7b.

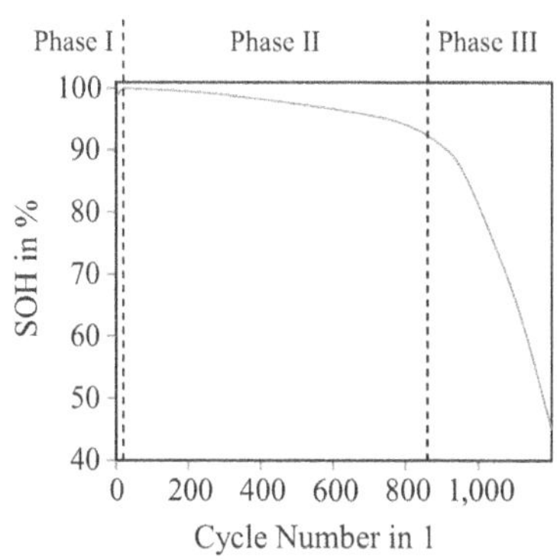

(a) Three aging phases

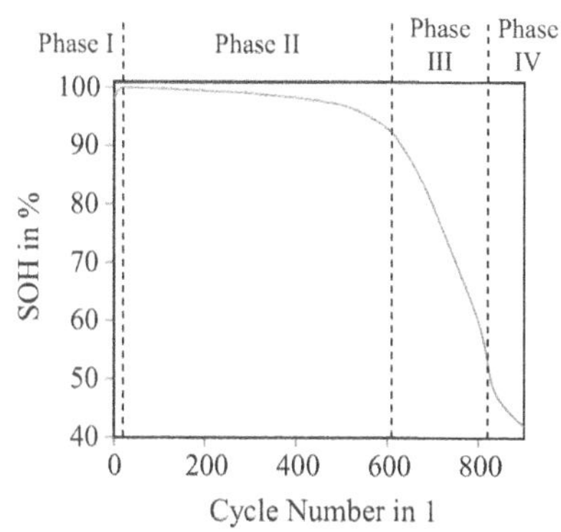

(b) Four aging phases

Figure 2.7 Aging phases of LIBs (based on [174, p. 2])

2.1.5 Battery Packs

In many applications including BEVs, not single battery cells, but battery packs are used. But most literature focuses on the SOH of battery cells [124; 179] as introduced as L3 in Chapter 1. Only few literature focuses on „full battery level“ [180], i.e., pack-level. In the literature survey in Chapter 4, only 20 % of the data sets are on pack-level (see Table C.2, accessible in the Electronic Supplementary Material).

A battery pack is an energy storage device composed of one or several electrically connected cells or modules [181; 182]. For ISO/IEC 62620:2014 [181] a battery pack optionally has a package housing as well as a protection unit for control and monitoring. The protection unit for control and monitoring is usually called BMS. In contrast to that, Kwade et al. [183] and Kampker [184] require the package housing, the BMS, and additionally a cooling system. Väyrynen et al. [182] do not mention these three components at all. Instead, they introduce the term battery system that includes one or multiple packs including a cooling system, the BMS' and peripherals. The terms battery pack and battery system often appear to be used as synonyms [185; 186] or even combined, e.g., as „battery-pack system" [187]. An overview of the discussed definitions is given in Table A.2 (accessible in the Electronic Supplementary Material). In our perception, the term battery pack emphasizes that the battery is physically packed somehow. The term battery system emphasizes that the battery is a system composed of several components. In this thesis, the definition of the term battery pack follows [181; 182]. The extension of the battery pack by housing, BMS, and cooling system, forms a battery system.

The BMS fulfills the following tasks [186; 188; 181, p. 3.11]:

1. **Data acquisition**: The basis for the other tasks is laid by acquiring data, e.g., about the cell voltages $V_{\mathrm{cell},i}$, pack voltage V_{pack}, the pack current I_{pack}, and several temperatures at different locations inside the battery pack $T_{\mathrm{pack},i}$.
2. **Battery state determination**: The battery states determined usually are the SOC and SOH.
3. **Electrical management**: The electrical management controls the charge and discharge process. This includes equalization charging, i.e., cell balancing, which ensures that the cells in a pack have the same SOC, i.e., voltage. Cell balancing maximizes the battery's usable amount of charge because the cell with the lowest voltage determines the end of discharging and the cell with the highest voltage determines the end of charging, both when hitting the respective voltage limits. Active cell balancing is only necessary in series connections. Due to Kirchhoff's voltage law, theoretically, cells in parallel connection balance each other automatically [49, p. 96] which however was disproved in practice with a retired BEV battery pack [189]. For further details on cell balancing consult [93, p. 426; 109, p. 14].
4. **Safety management**: The safety management protects the battery against critical operation like overcharging, overdischarging, high and low temperatures, as well as short circuiting.
5. **Thermal management**: The thermal management is cooling the battery and equalizing the temperature between cells.

6. **Communication**: In EVs, the BMS is communicating with other devices via the CAN bus.

A battery module is a group of battery cells in series (s) and/or parallel (p) connection [182–184; 186; 181, p. 3.8] as depicted in Figure 2.8. Usually, s and p circuits as depicted in Figures 2.8a and 2.8b respectively are combined: A p connection consisting of cells in s connection is called parallel-series (ps) connection as in Figure 2.8c. A s connection consisting of cells in p connection is called series-parallel (sp) connection as in Figure 2.8d. For example, 3s4p abbreviates three cells in s and four in p connection. The amount, connection, and split of the cells into modules and packs depends on the battery application. For further information consider [49; 184].[9]

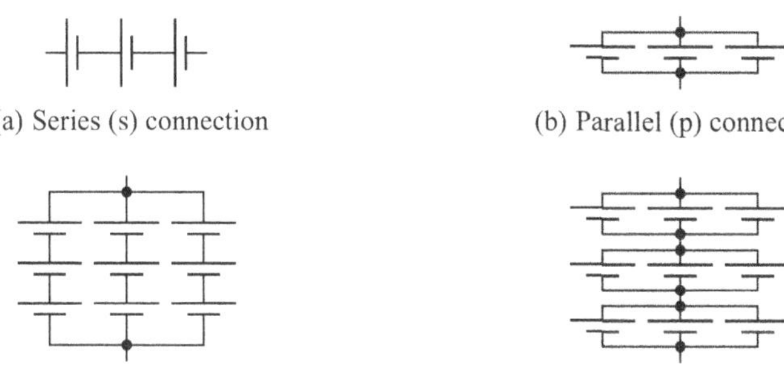

(a) Series (s) connection

(b) Parallel (p) connection

(c) Parallel-series (ps) connection (Also known as parallel strings)

(d) Series-parallel (sp) connection (Also known as strings of parallel)

Figure 2.8 Cells in series (s), parallel (p), parallel-series (ps) and series-parallel (sp) connection [194, p. 27–29; 195, p. 296; 191, p. 793]

2.1.5.1 States of Lithium-Ion Battery Packs

The capacity and SOH of a LIB pack depends on the LIB cells: Kampker [184, p. 57] states that the total performance of a battery pack is limited by the weakest cell in the pack. Warner [49, p. 172] extends this to the total performance and capacity

[9] Feng et al. [190] denote the combined connection topologies vice versa: The here depicted ps connection is denoted as sp connection by them and the sp vice versa as ps correspondingly. They describe their sp as series-connections in parallel and ps as parallel-connections in series. [191] call ps „parallel strings of cells" and sp „strings of parallel cells". [192; 193] use the terminology „parallel cell module" (PCM) for cells within a module wired in parallel and the modules wired in series (here sp). Correspondingly, a „series cell module" (SCM) contains cells wired in series and the modules wired in parallel (here ps).

being limited by the cell with the lowest capacity in the pack. More precisely, this only applies to cells in series connection because the cell with the lowest capacity will reach the voltage limits first [196; 197].

The capacity $C_s(t)$ of n_s cells in s connection is the lowest capacity of the cells in s connection $C_i(t)$ [194; 197]:

$$C_s(t) = \min_i C_i(t) \quad \text{with} \quad i = 1, 2, \ldots, n_s \tag{2.1.18}$$

assuming that all cells have the same internal resistance $R(t)$.[10]

The capacity $C_p(t)$ of n_p cells in p connection is the sum of the single cells' capacities $C_j(t)$ following Kirchhoff's current law [194; 198]:

$$C_p(t) = \sum_{j=1}^{n_p} C_j(t) \quad \text{with} \quad j = 1, 2, \ldots, n_p. \tag{2.1.19}$$

With Equations (2.1.14) and (2.1.18) follows, that the SOH_C of cells in s connection $SOH_{C,s}(t)$ is the cells' minimal $SOH_{C,i}(t)$ assuming C_{nom} is the same for all cells [194; 195; 199]:[11]

$$SOH_{C,s}(t) = \min_i SOH_{C,i}(t) \quad \text{with} \quad i = 1, 2, \ldots, n_s. \tag{2.1.20}$$

Analogous with Equations (2.1.14) and (2.1.19) follows that the SOH_C of cells in p connection $SOH_{C,p}(t)$ is the average of all single cells' $SOH_{C,j}(t)$ assuming C_{nom} is the same for all cells [194]:

$$SOH_{C,p}(t) = \frac{1}{n_p} \sum_{j=1}^{n_p} SOH_{C,j}(t) \quad \text{with} \quad j = 1, 2, \ldots, n_p. \tag{2.1.21}$$

[10] Assuming the cells are balanced well and, thus, the OCV is the same for all cells, still different R_i of cells in series connection lead to different voltage changes with the same current I_S on the series connection: $\Delta U_i = I_S \cdot R_i$. So, theoretically R_i also influences which cell hits the voltage limits first. Also, theoretically even for a cell with very high resistance the full capacity could be used with $\lim_{t \to \infty} I(t) \to 0$.

[11] The assumption that C_{nom} is the same for all cells is not made by Juhlin [194] but mathematically necessary. Its fulfillment is given in practice as the same battery cell type is usually built within one battery module or pack.

Thus, a higher degradation of a single cell has a bigger influence on the SOH_C in an s connection than in a p connection. This corresponds to the results of Jeng et al. [197]. Both Equations (2.1.20) and (2.1.21) are derived in Appendices A.2.1.2 and A.2.1.3 (accessible in the Electronic Supplementary Material). The same applies for the corresponding SOH_R in s connection and p connection in Appendices A.2.2.3 and A.2.2.4.

With Equations (2.1.20) and (2.1.21) follows that the $SOH_{C,ps}(t)$ of cells in ps connection is the average of all the p connections' SOH_C which each is determined by the minimum SOH_C of the s connection [194]:

$$SOH_{C,ps}(t) = \frac{1}{n_p} \sum_{j=1}^{n_p} \min_i SOH_{C,ij}(t) \tag{2.1.22}$$

with $i = 1, 2, \ldots, n_s$, and $j = 1, 2, \ldots, n_p$, and assuming C_{nom} is the same for all cells.

Analogous with Equations (2.1.20) and (2.1.21) follows that the $SOH_{C,sp}(t)$ of cells in sp connection is the minimum SOH_C of all s connections which each is the average SOH_C of the p connection [194]:

$$SOH_{C,sp}(t) = \frac{1}{n_p} \min_i \sum_{j=1}^{n_p} SOH_{C,ij}(t) \tag{2.1.23}$$

with $i = 1, 2, \ldots, n_s$, and $j = 1, 2, \ldots, n_p$, and assuming C_{nom} is the same for all cells. Both Equations (2.1.22) and (2.1.23) are derived step by step in Appendices A.2.1.4 and A.2.1.5 respectively (accessible in the Electronic Supplementary Material). Overall, it is desirable to build packs from cells with similar capacity. Also, the SOH of the same cells in sp connection is greater than or equal to the SOH of cells in ps connection:

$$SOH_{C,ps}(t) \leq SOH_{C,sp}(t) \tag{2.1.24}$$

as derived in Appendix A.2.1.6 and the Appendix of [2]. This is consistent with [192; 200].

2.1.5.2 Aging of Lithium-Ion Battery Packs

Aging of battery modules and systems is more complex than of a single cell because of the interactions of all cells. These interactions cause the degradation process of the battery pack and of the cells in that pack to depend on each other [74]. It is influenced by inconsistencies of cell characteristics, also known as intrinsic cell-to-cell variability [100], electrical imbalance, and temperature gradients between cells that cause heat transfer to adjacent cells [124; 201]. For example, Bruen et al. [202] observe significant differences of cells in parallel regarding their current which results in different temperature distributions and aging of the cells. Jung et al. [203] found that the internal temperature within battery modules is higher than of a single cell because of the heat transfer to adjacent cells. Tanim et al. [204] state that cell-level aging does not provide a direct foundation for extrapolation to pack-level aging. Calearo et al. [205] summarize that only few studies focus on the aging of the whole battery pack. However, cells in parallel strings with initially different SOH_C and SOH_R showed a convergence of the SOH difference [133]. Overall, Dubarry et al. [206] conclude that little effort has been made to study the different inhomogeneities and cell-to-cell variations in battery packs. For more information regarding the inconsistency of packs consider [207].

In this thesis, the main drivers of battery aging are assumed to depend on the three signals I, T, and SOC. The interdependencies of these signals when switching from cell to pack level are compared in Table 2.2.

Table 2.2 Interdependencies of I, T, and SOC battery aging when switching from cell to pack level

Signal	Cell level	Module and pack level
I	$I_{\text{cell},i}$	I_{pack} splits up to $I_{\text{cell},i}$ depending on series/parallel configuration and $R_{\text{cell},i}$ according to Kirchhoff's laws
T	Heat dissipation by cell i due to $I_{\text{cell},i}$ and $R_{\text{cell},i}$* and $T_{\text{environment}}$	Heat dissipation by cell i due to $I_{\text{cell},i}$ and $R_{\text{cell},i}$, heat transfer with adjacent cells, heat loss to cooling system,† and $T_{\text{environment}}$
SOC	$SOC_{\text{cell},i}$	SOC_{pack} depends on $SOC_{\text{cell},i}$ by the same logic as $C(t)$ and $SOH(t)$ considering the s/p configuration and cell balancing

* In laboratory operation of single cells usually no cooling system, but a heating chamber is used to accelerate aging. Convection is considered as possible, but radiation as negligible.
† In battery packs, radiation and convection are considered as negligible.

2.1.6 Battery Operation

Battery operation can show great differences like the operational context (laboratory vs. vehicle operation, see L1), different charging protocols, and potentially a cooling system. Discharging of LIBs in automotive applications depends on the driving style, traffic, and environment. Thus, there is only little potential to actively influence it by, e.g., a driver, fleet manager, or adaption of parameters of the BMS.

2.1.6.1 Charging of Lithium-ion Batteries

However, charging of LIBs in automotive applications depends on the available charging infrastructure, charging power, and the selected charging protocol. Among others, known charging protocols are: Constant current - constant voltage (CC-CV) charging, multistage constant current (MCC), pulsed charging, and boost charging (BC) [208].

Currently, the most common charging protocol for lithium-ion cells is CC-CV charging as illustrated in Figure 2.9. First, the battery is charged with a CC while the voltage and SOC increase. When the upper voltage limit has been reached, charging switches from CC to constant voltage (CV) charging. The voltage is kept constant by reducing the current so that the SOC increases more slowly. CV charging ends when the current falls below a certain small threshold [91, p. 73; 104, p. 446; 94, p. 92]. More information on charging protocols can be found in [208].

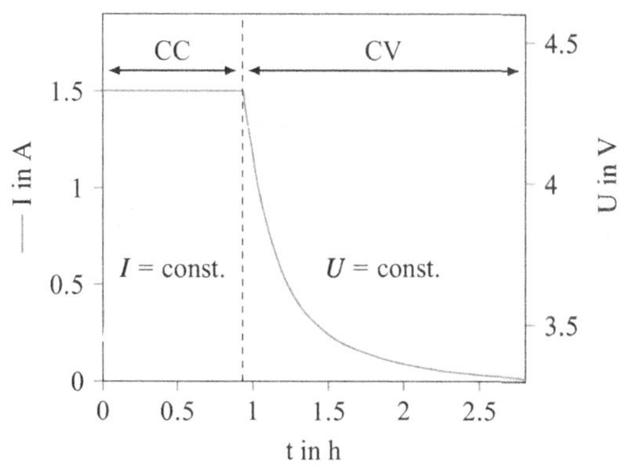

Figure 2.9 Schematic illustration of CC-CV charging

EVs can be externally charged using AC or direct current (DC). AC charging is using the vehicle's inverter to convert the, potentially three-phase, current to DC. The AC charger in the vehicle is limiting the charging power. Contrarily, with DC charging using the vehicle's inverter is not necessary as the charging station already converts and provides DC. This enables high powers. Still at roughly 80 % SOC the charging speed reduces when the CV charging phase is entered. Fast DC charging is known as high power charging (HPC) [208, p. 2; 209].

2.1.6.2 Differences of Battery Operation in Laboratory and Real-World Vehicle Operation

Battery operation in laboratory and real-world vehicle operation have differences that are relevant when analyzing and working with battery operational data [73; 210]. This also becomes relevant when key criterion 1c is introduced in Section 3.5.

When conducting battery aging experiments in the laboratory like in the public data sets presented by [68–71; 211] usually a test matrix or test procedure specifies the controlled operation conditions relevant for battery aging during the operation modes charging, discharging, and rest of the batteries. When considering a cycle of charging and discharging, the ΔSOC as well as possible recuperation need to be defined as well. Furthermore, during all three operation modes the ambient temperature is relevant. These criteria in Table 2.3 show that laboratory operation of batteries usually differs significantly from battery data from real-world operation of BEVs since test matrices cannot completely cover the real-world operating conditions [145; 212–214].

In laboratory operation, batteries are usually fully charged applying the CC-CV protocol [68; 71]. Some data sets also have multiple CC-steps before the CV-phase [69; 211]. In the real-world, a chosen charging protocol may be facing current drops, when additional BEVs start charging at the same charging station and power supply is limited [215]. Also, the charging management may limit the charging

Table 2.3 Differences of laboratory and real-world operation of batteries

Aging	Criterion	Laboratory operation	Real-world BEV operation
Cyclic	Charging	• CC-CV or • multiple CC-steps, then CV	Similar to laboratory, but variable duration, current drops, transient load, and possible current limitations
	Discharging	CC or Randomized	Highly variable current
	Recuperation	None	Yes
	ΔSOC	100 % or other constant value	Highly variable
Calendar	Hold/Rest/ Storage	• Non or • Only short or • Solely for calendar aging experiments	Dominant operation mode
Cyclic & calendar	Temperature	• Constant room temperature or • Temperature chamber without cooling system	Variable, depending on season, but with cooling system

current depending on the batter's temperature. Furthermore, BEV batteries are not always fully charged. Especially, when applying fast-charging at highways the CV-phase is usually skipped [216].

In laboratory operation, batteries are usually discharged using a single CC-step [69–71; 211]. Only one public data set applies multiple CC-steps of 5 minutes length with randomized discharge currents [68]. However, in real-world automotive applications the discharge current is not constant, but highly variable, non-linear and contains huge, but brief power demands [73; 215; 217; 218]. Further differentiation regarding the current profile is given later in this section.

Due to regenerative braking, i.e., recuperation, in automotive applications charging and discharging mode alternate frequently, especially when driving in urban areas [217]. From a battery point of view, recuperation can be seen as micro-charging during a longer discharging process which is also called partial or micro cycling. This is usually not considered by test matrices of laboratory operation.

Fully charging and discharging, i.e., $\Delta SOC = 100\ \%$ is common in laboratory operation [69–71]. Alternatively, the SOC_{start} and SOC_{end} of each cycle are constant so that $\Delta SOC = \text{const.}$ [211]. However, in real-world applications, ΔSOC of cycles are highly variable [216].

Furthermore, BEVs usually do not use low SOCs frequently due to the range anxiety of the drivers [48; 219]. However, in HEVs the complete range of the SOC is used. The same applies for BESS. These can be used for the optimization of own consumption (private and commercial), peak-shaving, and frequency regulation (primary control) [220]. Corresponding markets are the frequency containment reserve (FCR), intraday continuous market, and day-ahead auction market [221]. Compared to automotive HV battery applications, batteries in BESS applications have around 3 full cycles per day [222; 223].

If calendar aging is not examined in laboratory operation, rest phases without any current applied are usually temporally negligible compared to the duration of charging and discharging phases. Usually, they are only part of the experiments, e.g., for voltage relaxation and characterization tests. However, in standard automotive and BEV use cases rest phases, i.e., inactive HV system and open HV contactors, are usually dominant [217; 224]. Even if the BEV is not moving, some current is always applied for auxiliary consumers, like the air conditioning or the entertainment system. Only in MOD use cases of BEVs rest phase are less dominant because the vehicles are operated more intensively [225].

In laboratory operation batteries are usually exposed to a constant ambient temperature which is either the room temperature or forced by a heating/cooling chamber [226]. In contrast, batteries operated in automotive applications face the complete seasonal range of temperatures from a cold winter night to a hot summer noon, especially when parked outside. Compared to automotive applications, batteries operated in BESS usually have better thermal environment conditions.

Current profile: From Laboratory to Real-World Vehicle Battery Operation Furthermore, battery operation can be differentiated by the current profile applied to the battery as structured by the 2x2 matrix in Table 2.4. The current profile is the abstracted time series signal of the current I or C-rate applied to a battery for charging and discharging. The abstracted time series does not consider any small variation of the current due to current control, e.g., during CC and CV phases. During external charging, i.e., excluding brake energy recuperation during driving, usually charging protocols are used so that the differences of the current profiles of laboratory and real-world operation usually originate from the discharging operation. Petrovic [102] distinguishes three modes of discharge: CC, constant load, and constant resistance. However, as he notes, automotive applications face variable discharging.

Two criteria to differentiate battery aging data sets are proposed: First, the current profile can be distinguished as identical (I) over all cycles, usually for laboratory operation, or as non-identical (NI) over all cycles (Rows of Table 2.4). Second, the current profile can be distinguished as CC or variable current (VC) as in the columns of Table 2.4.[12] When having CC, also optional short CV phases are included, i.e., non-CC shall not be caused by a controller during a VC phase.

Thus, an identical, single CC step and a optional CV phase define the 1st type of I-CC (I-CC.1). The 2nd type of I-CC has identical, MCC steps (I-CC.2). These two are common types of charging protocols [208]. The exact time length of the CC steps may vary due to battery aging.

The next step towards real-world vehicle operation is still with CC steps, but the CC steps of different cycles are NI (NI-CC). This may be the case if the current of each CC step is random sampled such as in [68]. The random CC sampling better represents practical battery usage than non-random CC operation [208].

Further, with VC still the current profiles of different cycles can be identical (I-VC), as first, the current profile can simply be a replay of a recorded profile, e.g., from vehicle operation that is repeated for each cycle (I-VC.1). Second, the current profile can be defined by a representative profile (I-VC.2). Examples are the Worldwide harmonized Light vehicles Test Procedure (WLTP)[13] [241], Urban

[12] These criteria are also called „change during a single cycle“ (CC vs. VC) and „change between different cycles“ (I vs. NI) by [240].

[13] The older New European Driving Cycle (NEDC) (German: Neuer Europäischen Fahrzyklus (NEFZ)) only consists of linear slopes and constant phases of velocity. Thus, its current profile will be mostly constant.

Table 2.4 Characterization of battery aging data sets in a 2x2 matrix by current profile applied to the battery. Referenced papers present or use data sets of the according category

		Current		
		Constant current (CC) steps (optional CV)	Variable current (VC)	
Current profile over cycles	Identical (I)	I-CC: 1) 1 CC step [71; 169; 211] 2) Multistage constant current (MCC) [69; 70]	I-VC: 1) Replay of recorded current profile from vehicle [145; 227] 2) Representative profile [227–234] 3) Driver following predefined profile [235]	→ laboratory to real-world battery operation →
	Non-*identical* (NI)	NI-CC: 1) *CC steps with* randomized current 68; 236	NI-VC: 1) *Real-world vehicle* operation [237–239]	
→ From laboratory to real-world vehicle battery operation →				

Dynamometer Driving Schedule (UDDS) [242] or any representative driving cycle (RDC) that aggregately represents real-world operation [243]. Third, a unique professional driver can reproduce a certain current profile on a private track as in [235] (I-VC.3). In this case, the current profiles can only approximately be identical over multiple cycles.

Lastly, NI current profile over multiple cycles and VC (NI-VC.1) specifies real-world vehicle operation in which no cycle-wise exact current patterns can be found. This is caused by different drivers, routes, traffic, weather, etc.

Summarized, there is a transition of battery aging data sets from strong laboratory conditions to real-world vehicle operation from I-CC via N-ICC and I-VC to NI-VC. However, most public battery aging data sets are of the category I-CC [72].

Further criteria of distinction are mentioned in Table 2.5: First, the place of battery operation will likely be a test bench for I-CC, I-VC and NI-CC except I-VC.3 or alternatively a vehicle for I-VC.3 and NI-VC.1. Second, most data sets from test benches either have cyclic or calendar aging. Data sets applying cyclic and calendar aging to the batteries are rare [169]. However, in real-world vehicle operation NI-VC, this is normal with dominating calendar aging [2]. Third, also for real-world vehicle operation NI-VC, normally batteries in vehicles have a cooling system. This is usually not the case for operation on a test bench. Fourth, the battery environment temperature on test benches is usually either a nearly constant room temperature or defined by a heating chamber. In real-world vehicle operation NI-VC, the batteries face the outdoor temperature with seasonal changes. Fifth, most battery aging data sets use single cells. Battery modules are rarely used on test benches and only in small quantities as in [244][14] (I-CC.1) and [245][15] (I-VC.1). Battery packs are built in vehicles (I-VC.3 and NI-VC.1), but when such data sets are used, e.g., by [237; 238], they have not been published, probably due to confidentially.

[14] [244] provide data of 4 packs each with 11 NMC 40Ah cells (4s11p) operated in series connection.

[15] [245] provide data of 2 packs each with 16 cells (16s1p) operated for 20 and 200 cycles respectively.

Table 2.5 Further criteria to distinguish battery aging data sets

No.	Criterion	Possible Value
1	Place of battery operation	Test bench or in-vehicle operation
2	Aging categories	Cyclic or calendar aging
3	Cooling system	Yes or no
4	Environment temperature	Room temperature, heating chamber, or outdoor temperature with seasonal changes
5	Battery type	Cell, module, or pack

2.2 Supervised Machine Learning for Regression Problems

Given the importance of ML for SOH forecasting, this chapter provides information on important concepts used in the context of artificial neural networks (ANNs) like feedforward neural networks (FNNs), recurrent neural networks (RNNs), training algorithm, feature scaling and hyperparameter tuning. Also, the ML paradigm of transfer learning is presented.

2.2.1 Artificial Neural Networks

ANNs emerged from attempts to mathematically represent information processing in biological systems like the brain [246]. They are loosely inspired by modeling biological neural networks. Today ANNs are successfully applied in function approximation as well as several other tasks such as pattern, image, and speech recognition [247, p. 167]. Just like in the brain, ANNs are composed of interconnected neurons. In ANNs, these are structured in layers. From known inputs an ANN calculates outputs based on its internal structure of layers. Two types of ANNs can be distinguished: FNNs and RNNs.

2.2.1.1 Feedforward Neural Networks

Multilayer perceptrons (MLPs) are a type of feedforward neural networks (FNNs). According to the universal approximation theorem, an MLP with a single hidden layer can approximate any smooth function from one finite dimensional space to another to any desired degree of accuracy as long as sufficient hidden neurons are provided [248]. However, no statement is made about the optimality of this approximation regarding the learning time, the ease of implementation or generalization [249]. Still, they are suitable for regression tasks.

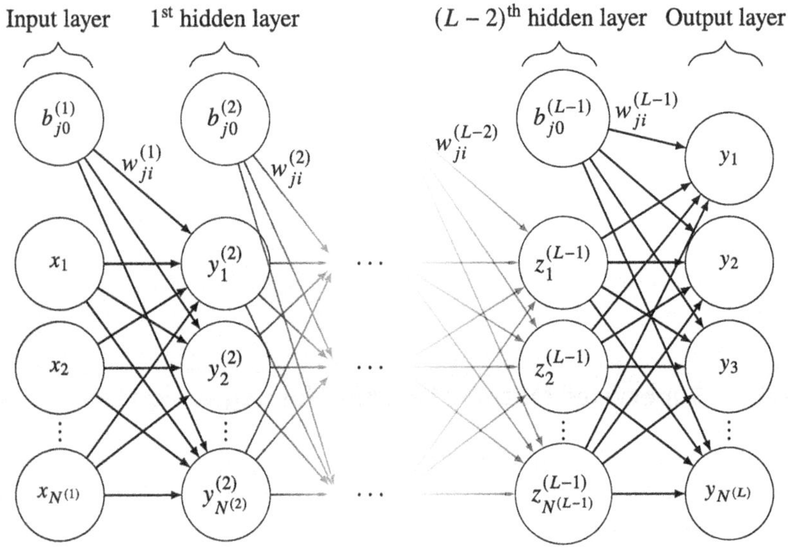

Figure 2.10 Example of an MLP with $N^{(1)}$ inputs, hidden layers with $N^{(2)}$ and $N^{(L-1)}$ neurons and $N^{(L)}$ outputs

As depicted in Figure 2.10, the MLP can be described as a directed acyclic graph connecting L layers, numbered from $l = 1$ to L. The input layer $l = 1$ is imaginary and corresponds to the inputs of the MLP. The last layer $l = L$ corresponds to the outputs. Between these two layers there may be several hidden layers $l = 2, \dots, L-1$ each consisting of $N^{(l)}$ neurons, but at least one hidden layer is necessary. Except of the input layer, i.e., for all $l \geq 2$, the value $z_j^{(l)}$ of the neuron j in the layer l depends on the values of the neurons $z_j^{(l-1)}$ in the previous layer $l-1$ with the weights $w_{ji}^{(l-1)}$ and biases $b_{j0}^{(l-1)}$ given:

$$z_j^{(l)} = h\left(\underbrace{\sum_{i=1}^{N^{(l-1)}} w_{ji}^{(l-1)} \cdot z_i^{(l-1)} + b_{j0}^{(l-1)}}_{=:a_j^{(l)}}\right) \tag{2.2.1}$$

The weight $w_{ji}^{(l-1)}$ describes the connection strength of the neuron i of the previous layer $l-1$ to the neuron j of the next layer l. The bias $b_{j0}^{(l-1)}$ is independent of the input $z_i^{(l-1)}$ and useful for including any fixed offset in the data like unforeseen or non-observable factors.[16] The weighted sum of all inputs and the bias $a_j^{(l)}$ is used as argument of the nonlinear activation function $h : \mathbb{R} \rightarrow \mathbb{R}$ to calculate the value $z_j^{(l)}$ of the neuron j in the next layer l. Common activation functions are among others the hyperbolic tangent (tanh) and the rectified linear unit (ReLU) [250; 251]. Equation (2.2.1) describes the information processing of a single neuron in the network during forward propagation to calculate the predicted output $\hat{\mathbf{y}}$ of an MLP for a given input vector $\mathbf{x_n}$.

2.2.1.2 Recurrent Neural Networks

Recurrent neural networks (RNNs) are a group of ANNs that process sequential data such as textual data or time series data, e.g., battery time series data from the CAN bus. An MLP has separate parameters for each input feature. In contrast, Recurrent neural networks (RNNs) recursively share the same parameters θ over several time steps. Thus, when unfolding this recursive, i.e., recurrent, structure, the current state $\mathbf{h^{(t)}}$ depends on the previous state $\mathbf{h^{(t-1)}}$, the current input $\mathbf{x^{(t)}}$, and the parameters θ of the RNN:

$$\mathbf{h^{(t)}} = f(\mathbf{h^{(t-1)}}, \mathbf{x^{(t)}}, \theta). \tag{2.2.2}$$

The RNN learns to keep information from past inputs in the current state [247]. Given their recurrent structure, RNNs have problems of learning long-term dependencies because of vanishing or exploding gradients [247; 252; 253]. From here on different types of RNNs can be distinguished like gated recurrent unit (GRU) [254] and long short-term memory (LSTM) [255].

2.2.1.3 Training Algorithms

Training an ANN is the process of minimizing a loss function $J(\mathbf{W}, \mathbf{b})$, also referred to as cost or error function, by finding optimal parameters $\mathbf{W}$ and $\mathbf{b}$, respectively weights and biases. The training of ANNs can be perceived as an optimization with weights and biases of the network as adjustable parameters. For the supervised training a data set $\mathbf{S} = (\mathbf{x_n}, \mathbf{y_n})$ of input data points $\mathbf{x_n} \in X \subset \mathbb{R}^m$ and output data points $\mathbf{y_n} \in Y \subset \mathbb{R}^o$ with $n = 1, 2, \ldots, N$ is given where m and o specify the dimension of the input and output vector respectively.

[16] The biases are sometimes also denoted as $b_j^{(l-1)}$ or $w_{j0}^{(l-1)}$.

The cost function measures the deviation of all N pairs of target output in the data set $\mathbf{y_n}$ and predicted output by the ANN $\hat{\mathbf{y}}_\mathbf{n}(\mathbf{x_n};\mathbf{W}, \mathbf{b})$ parameterized by $\mathbf{W}$ and $\mathbf{b}$. A common cost function for regression problems is defined in Equation (2.2.6):

$$\min_{\mathbf{W},\mathbf{b}} J(\mathbf{W}, \mathbf{b}) = \frac{1}{2N} \sum_{n=1}^{N} \underbrace{\left\| \mathbf{y_n} - \hat{\mathbf{y}}_\mathbf{n}(\mathbf{x_n}; \mathbf{W}, \mathbf{b}) \right\|^2}_{=J_n(\mathbf{W},\mathbf{b})} = \frac{1}{2N} \sum_{n=1}^{N} \sum_{o=1}^{O} (\mathbf{y}_{\mathbf{n}_o} - \hat{\mathbf{y}}_\mathbf{n}(\mathbf{x_n}; \mathbf{W}, \mathbf{b})_o)^2, \tag{2.2.3}$$

where $\mathbf{W}$ represents a tensor with $L - 1$ elements containing the weight matrices $\mathbf{W}^{(l)}$ for each layer. Each weight matrix consists of the elements $w_{ji}^{(l)}$. The sums are over all N data points and all O outputs of the ANN. A factor of $\frac{1}{2}$ is added to the mean squared error (MSE) to eliminate the factor 2 in the derivative of the cost function that originates from the power of 2 of the mean square error (MSE). It is important to note that the cost function in Equation (2.2.6) is based on the MSE which can also be used as a metric to evaluate the performance of the ANN on a given data set as explained in Section 2.2.4 [247; 256].

2.2.1.4 Epochs and Batch Size

When the training data set $\mathbf{S}_{\text{train}}$ is too big to load it all at once into the memory, the training data need to be divided into smaller data sets (batches). These batches are given to training algorithm one by one and after each one of them the weights and biases of the neural network are updated.

An epoch is an iteration over the entire training data set provided. It is usually chosen large to allow the training algorithm to run until the error of the MLP has been minimized sufficiently.

The batch size defines the number of samples to be used per gradient update. The batch size is typically between 1 and a few hundreds. Increasing the batch size takes advantage of the speedup of matrix-matrix products over matrix-vector products. A small batch size achieves best training stability and generalization performance. Thus, an intermediate batch size balancing these two effects is best. If the data set is not divided evenly by the batch size, the last batch simply has fewer samples than the other batches [257, pp. 442–443; 258; 247, pp. 152–153].

Depending on the batch size b_{size} three variants of gradient descent can be distinguished:

- Stochastic gradient descent: $b_{\text{size}} = 1$,
- Mini-batch Gradient descent: $1 < b_{\text{size}} < N_{\text{train}}$,
- Batch gradient descent: $b_{\text{size}} = N_{\text{train}}$,

where N_{train} is the size of the training data set $\mathbf{S}_{\text{train}}$. More information can be obtained from [259].

2.2.2 Feature Scaling

Feature scaling is used to standardize the range of the inputs. Usually, the raw data set contains input values which vary highly in the order of magnitude, range,[17] and unit. Since most ANN training algorithms apply the Euclidean distance in the loss function, one input variable with a wide span of values will dominate the Euclidean distance and prevent the ANN to learn from other input variables correctly. Hence, the range of all input variables should be scaled so that each input variable contributes equally to the total Euclidean distance. The same problem arises in the case of different unit prefixes. Feature scaling is usually executed during the preprocessing of the data. Furthermore, gradient descent algorithms converge much faster with feature scaling applied [260; 261].

The most prominent feature scaling methods are standardization and rescaling which are presented in the succeeding paragraphs. In the following, $\mathbf{x}$ is the vector containing the original values of one input variable, $\mathbf{x}_{\text{scaled}}$ is the vector containing the scaled values of one input variable, μ is the mean and σ the standard deviation (SD) of the input variable.

Standardization is widely used for normalization in many ML algorithms and also known as Z-score normalization. It redistributes the features with their new mean $\mu_{\text{scaled}} = 0$ and SD $\sigma_{\text{scaled}} = 1$ [262]:

$$\mathbf{x}_{\text{scaled}} = \frac{\mathbf{x} - \mu}{\sigma} \tag{2.2.4}$$

with the addition and division applied component-wise [260].

Rescaling is the simplest method and useful for dealing with input variables with hard boundaries. It is also known as min-max scaling or min-max normalization because it rescales the range of the input variable to the range $[0, 1]$ or $[-1, 1]$:

[17] E.g., $x_1 \in [50, 100]$ has the same range like $x_2 \in [4.950, 5.000]$, but a different order of magnitude. Also, $x_1 \in [90, 100]$ has the same order of magnitude like $x_2 \in [99, 100]$, but a different range.

$$\mathbf{x}_{\text{scaled}} = \frac{\mathbf{x} - \min(\mathbf{x})}{\max(\mathbf{x}) - \min(\mathbf{x})} \tag{2.2.5}$$

with the addition and division applied component-wise and for the elements $x_{\text{scaled},i} \in [0, 1]$.

The selected range depends on the input data. Rescaling decreases the SD, which can suppress the effect of outliers [261].

2.2.3 Hyperparameter Tuning

Hyperparameters are parameters of the training algorithm and not internal model parameters found by the training algorithm like weights and biases. Part of training an ANN is choosing a set of optimal hyperparameters that must be determined before training commences, shortly called hyperparameter tuning or hyperparameter optimization. The hyperparameter choice significantly affects the performance of the resulting model. However, determining good hyperparameter values in advance can be a complex task.

The hyperparameters that are included in this search process depend on the ANN structure. Typical hyperparameters are the learning rate, the chosen training algorithm, the batch size, number of epochs to be trained, the regularization parameter, and the dropout rate [263].

In the following, grid search, random search, and tree-structured parzen estimator (TPE) as hyperparameter optimization algorithms are presented. Another algorithm for hyperparameter optimization is Bayesian Optimization [264; 265].

Grid Search exhaustively generates candidates from a manually defined grid of hyperparameters, which contains reasonable values for each hyperparameter. Each combination of the hyperparameters in the grid is used as a hyperparameter set to train a model. After each training, the model's performance is evaluated on a held-out validation set. Finally, grid search selects the best hyperparameter set in the grid space. On the one hand, grid search is limited due to the curse of dimensionality because all possible combinations of hyperparameters are used for training. On the other hand, grid search is parallelizable since the hyperparameter sets can be used for training and evaluation independently of each other [266].

Random Search selects a random combination of values for each hyperparameter. In contrast to grid search, the hyperparameter space does not require discretization. Furthermore, compared to grid search, random search can be stopped at any time and the result will still be somehow complete as the hyperparameter space has evenly be discovered. However, both grid search and random search are non-adaptive, i.e., they do not consider any knowledge obtained about the hyperparameter space during the search [266].

Tree-structured Parzen Estimator is a sequential model-based global optimization (SMBO) algorithm. SMBO is useful when the evaluation of a certain fitness or loss function is expensive. In the case of hyperparameter tuning, this function is the performance of the ML model on the validation data set with the model's hyperparameters as independent variables. SMBO approximates this function by a surrogate model with cheaper evaluation. In SMBO, this surrogate model is iteratively used to predict the performance of a certain hyperparameter combination and then to fit the real performance of this hyperparameter combination. Because of the consideration of the results of previous hyperparameter combinations SMBO achieves better results at cheaper cost compared to grid search and random search [267–269]

2.2.4 Metrics for Regression Problems

The performance of ML models on a given data set can be evaluated by different metrics. Some of these metrics like the MSE, root mean squared error (RMSE), or mean absolute error (MAE) are also used in the loss function.

The **mean squared error (MSE)** is the mean squared deviation of the true output value y_i and the predicted value of the model $\hat{y}_i$:

$$MSE = \frac{1}{N}\sum_{i=1}^{N}(y_i - \hat{y}_i)^2. \tag{2.2.6}$$

By squaring the deviation of y_i and $\hat{y}_i$, larger errors weigh in more than smaller ones. Following, outliers in the error increase the MSE more. A disadvantage that results from squaring the deviation is a squared unit, making the direct interpretability more difficult [270; 271].

Taking the root of the MSE results in the **root mean squared error (RMSE)** whose unit is not squared anymore:

$$RMSE = \sqrt{MSE} = \sqrt{\frac{1}{N}\sum_{i=1}^{N}(y_i - \hat{y}_i)^2}. \tag{2.2.7}$$

The **coefficient of determination (R^2)** represents the percentage of variance in the outputs y_i explained by the model. The highest score possible is 1.0. However, overfitting may occur when a perfect regression model for the training data set performs poorly on the validation and test sets [272; 273]. A model that predicts always the mean $\overline{y}$ results in coefficient of determination (R^2) of 0.0. A negative R^2 means the model predicts worse than always predicting the mean $\overline{y}$ as shown in Equation (B.4) (accessible in the Electronic Supplementary Material). In other words, R^2 normalizes the MSE by the variance σ_y^2 of all output values y_i:

$$R^2 = 1 - \frac{MSE}{\sigma_y^2} = 1 - \frac{RMSE^2}{\sigma_y^2}. \tag{2.2.8}$$

The range of MSE and RMSE is $[0, \infty)$. These metrics do not express any information about the model's performance with respect to the distribution of the y_i. R^2 however does so. Since σ_y^2 is constant for a certain data set, R^2 is negative monotonically related to MSE and root mean square error (RMSE). This means ordering model's on the basis of MSE has the reverse order compared to ordering by MSE and RMSE [274].

Further information on R^2 is given in Appendix B.1 (accessible in the Electronic Supplementary Material).

2.2.5 Transfer Learning

In the field of battery models, transfer learning is often used to adapt an existing ML model to a new domain like a new battery type or different operational conditions. Transfer learning is a learning paradigm in ML that utilizes knowledge previously attained in one domain to solve a task in a novel domain [275].

A domain $D = \{\mathcal{X}, P(X)\}$ is defined by a feature space $\mathcal{X}$ and a marginal probability distribution $P(X)$, where $X = \{x_1, \ldots, x_n\}$, $x_i \in \mathcal{X}$ with n samples. X represents a particular sample that is made up of different observations x_i which all lie in the feature space $\mathcal{X}$.

A task $T = \{\mathcal{Y}, f(\cdot)\}$ is defined by a label space $\mathcal{Y}$ and an objective predictive function $f(\cdot)$, which is learned from the training data. $f(\cdot)$ can be seen as $P = (y|x)$ from a probabilistic perspective

In general, source domain data is assumed as $D_S = \{(x_{S_1}, y_{S_1}), \ldots, (x_{S_n}, y_{S_n})\}$, where $x_{S_i} \in \mathcal{X}_S$ is an input vector and $y_{S_i} \in \mathcal{Y}_S$ is the corresponding output vector. The target domain data can be denoted as $D_T = \{(x_{T_1}, y_{T_1})), \ldots, (x_{T_n}, y_{T_n})\}$, where $x_{T_i} \in \mathcal{X}_T$ is an input vector and $y_{T_i} \in \mathcal{Y}_T$ is the corresponding output vector [275].

Using these formal definitions, transfer learning can be defined in the following way:

„Given a source domain D_S and learning task T_S, target domain D_T and learning task T_T, transfer learning aims to help improve the learning of the target predictive function $f(\cdot)$ in D_T using the knowledge in D_S and T_S, where $D_S \neq D_T$, or $T_S \neq T_T$." [275]

When applying transfer learning, Torrey et al. [276] identify three measures by which learning might be improved due to a transfer. These are visualized in Figure 2.11 at the exemplary model performance of a regression task measured by the RMSE:

Start, convergence speed and the asymptote of the model's performance may improve. First, the initial RMSE before any training with data from the target domain might be lower with transfer learning than without. This occurs when a model, i.e., an ANN trained on source domain data provides a better start for learning than a randomly initialized model. Second, transfer learning can increase the convergence speed, i.e., it can decrease the training time to fully learn the target task. Third, the final performance achievable on the target task with transfer learning can be better than without transfer learning. The lower start and lower asymptote can easily be measured by the model performance at the start and end of training respectively. However, measuring the higher convergence speed is more difficult. A popular metric is the area under the learning curve (AULC). For the convergence speed the improvement of the model performance is interesting. Thus, the exponent of a power-law fit is more suitable [277; 278]. Mathematically, this measure is concerning relative convexity [279].

Further, Pan et al. [275] name three main concerns relevant when implementing transfer learning: What, how and when to transfer. What to transfer concerns the part of knowledge that is transferable across domains or tasks. On the one hand, the knowledge may be domain or task specific. On the other hand, it may be common across different domains and tasks. In the latter situation transfer learning may be beneficial. How to transfer concerns the learning algorithm for transfer learning, such as fine tuning, layer-wise freezing of an ANN or domain adaptation. Lastly, when to transfer also includes the question of when not to transfer to avoid a negative impact on the performance in the target domain which is known as negative transfer.

Transfer learning is well established in the field of computer vision. For example, for image classification pretrained convolutional neural networks (CNNs) like VGG-16, ResNet50, and Inceptionv3 are popular. These models were trained on large data sets, like the VGG-16 on 1.3 million images with 1000 classes [280]. Yosinski et al. [281] state that on the first convolutional layers CNNs learn abstract features such as edges and corners that are relevant for many data sets and tasks. When

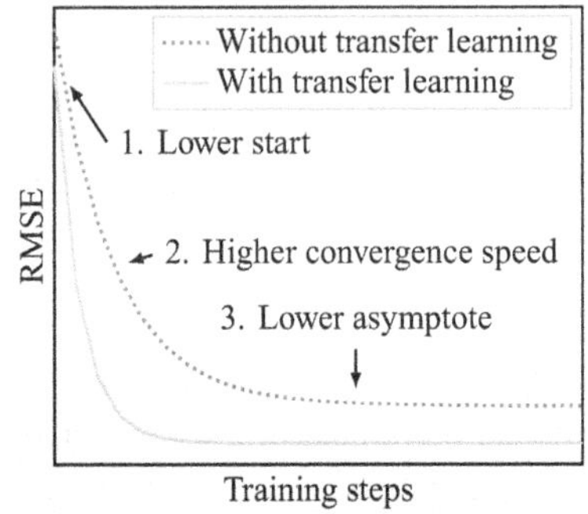

Figure 2.11 Measures in which transfer might improve learning: 1. Lower start, 2. higher convergence speed, and 3. lower asymptote (adapted from [276, p. 2])

only a much smaller target data set is available, learning such abstract features with a good generalization might be difficult. In this case, the first layers of the pretrained CNN are useful as feature extractor and improve generalization. Building on these features, some of the higher convolutional layers and the classifier part of the CNN generate specialized features [282]. The major advantage of transfer learning in computer vision is the elimination of the lengthy training process and improved generalization. Rawat et al. [283] provided a comprehensive overview of the application of CNNs to visual tasks.

Towards State of Health Forecasting of Lithium-Ion Batteries

3

Building on the theoretical background laid in the previous Sections 2.1 and 2.2, the terms SOH estimation, SOH prediction, SOH forecasting, and RUL prediction are crucial to distinguish for this thesis. However, these terms are used inconsistently in the literature and their conceptualization is complex and vague until now. Thus, this chapter introduces a formal concept and definitions mainly for SOH estimation and SOH forecasting. Furthermore, the relevance of model transfer for SOH forecasting to a new battery type is motivated in Section 3.4. This chapter also delimits this thesis' object of investigation so that Section 3.5 motivates key criteria SOH forecasting models shall fulfill.

This chapter is based on previous publications [1; 2].

3.1 State of Health Estimation, Prediction, and Forecasting

In the literature, two methodological principles appear: SOH estimation and SOH forecasting.[1] Their distinguishing characteristics are summarized in Table 3.1.

[1] [72, Fig. 5] distinguish both as "short term SOH estimation" and "future trajectory prediction" respectively.

F. von Bülow, *A Data-Driven Fleet Service: State of Health Forecasting of Lithium-Ion Batteries*, AutoUni – Schriftenreihe 170,
https://doi.org/10.1007/978-3-658-43188-4_3

Table 3.1 Distinguishing characteristics of SOH estimation and forecast

Distinguishing characteristic		Estimation	Forecast
Modeling		State determination x	State change $\dot{x}$ or Δx
Time window τ	Length	Short	Long
	Relative temporal position to t_1	In past	In future
$\mathbf{X}_\tau = SOC_\tau, I_\tau, T_\tau, \ldots$	Features encode	Electrochemical behavior	Applied load
	Features encode ... of battery aging	Effect	Causes

First, **SOH estimation** is the determination of the $SOH(t_e)$ at a certain point in time t_e based on information available at this point in time, as illustrated in Figure 3.1a:

$$SOH(t_e) = f_e(\mathbf{X}_\tau) \tag{3.1.1}$$

with the function's argument $\mathbf{X}_{\tau_e} = SOC_{\tau_e}, I_{\tau_e}, T_{\tau_e}, V_{\tau_e}, \ldots$ as time series of SOC, current I, temperature T, and voltage V in the time window $\tau_e = [t_{past}, t_e]$ that can either comprise a few cycles [52], a single cycle [54] or less [284].[2] Most importantly τ_e shall not cover any significant SOH change over time. The point in time t_e may be just now or in the past ($t_e \leq t_1$ with t_1 as current point in time). The electrochemical behavior expressed in the signals of SOC, I, T, and V changes due to battery aging [288]. Hence, SOH estimation is based on features that reflect the effect of battery aging (see right in Figure 2.5).

Second, the basis of **SOH forecasting** is the perception of battery aging as an SOH state change $\Delta SOH(\tau_f)$ during the time period $\tau_f = [t_1, t_2]$ from a current $SOH(t_1)$ to a future $SOH(t_2)$ due to aging causes given the battery operational load during the time period τ_f which is also called forecast horizon [212; 289]. Thus, a 1st possible forecast function $f_{f,I}$ is:

$$\Delta SOH(\tau_f) = f_{f,I}(SOH(t_1), \mathbf{X}_{\tau_f}) \tag{3.1.2}$$

with the arguments $\mathbf{X}_{\tau_f} = SOC_{\tau_f}, I_{\tau_f}, T_{\tau_f}, \ldots$ as time series of SOC, I, T, and V in the time window τ_f that express the causes of the battery aging due to battery operational load during the time period τ_f explicitly, as illustrated in Figure 3.1b (see left in Figure 2.5). This ensures that the battery operational load is not implicitly

[2] For further information consider reviews on SOH estimation [99; 11; 122; 124; 285–287].

considered to be fixed [150] or derived from the past load ($t \leq t_1$) as introduced as L2 in Chapter 1. As batteries age over a long time, τ_f is longer than τ_e. Also, the derived features may be different, but they are derived from the same time series, i.e., raw data, as Table 3.1 indicates.

Alternatively to the future change $\Delta SOH(\tau_f)$ by $f_{f,I}$, also the future $SOH(t_2)$ can be forecasted directly by a 2nd possible forecast function $f_{f,II}$:

$$SOH(t_2) = f_{f,II}(SOH(t_1), \mathbf{X}_{\tau_f}) \tag{3.1.3}$$

with the functions $f_{f,I}$ and $f_{f,II}$ being related by

$$f_{f,II}(SOH(t_1), \mathbf{X}_{\tau_f}) = SOH(t_1) + f_{f,I}(SOH(t_1), \mathbf{X}_{\tau_f}) \tag{3.1.4}$$

while learning the functions $f_{f,I}$ or $f_{f,II}$ in a regression model enables SOH forecasting. Liu et al. [290] also called this "future capacity prognostics".

In the literature, apart from SOH estimation and SOH forecasting also the term SOH prediction is often used which seems ambiguous to SOH forecasting. In general, the terms "forecast" and "prediction" are often used synonymously in practice. However, dictionaries show slight differences in their definition [291–296], just as Petropoulos et al. [297] describe forecasting as prediction about the future. Chugh [298] and Döring [299] define forecast as prediction with temporal dimension.

When having an SOH estimation model created based on past and present data, using this model to estimate a future SOH with unseen data from the future, this is called SOH prediction [299]. Hence, using SOH estimation models for prediction

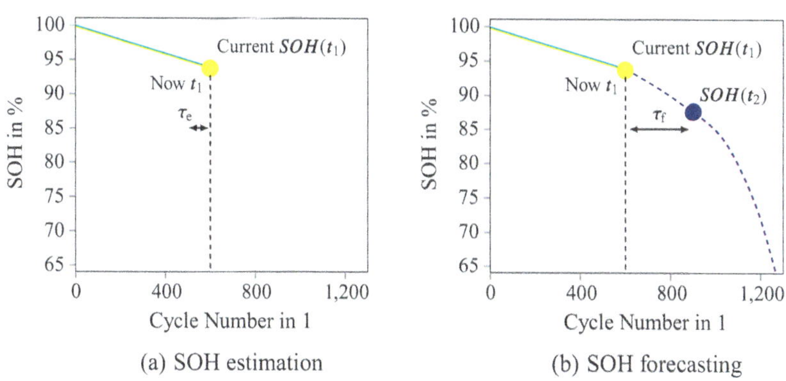

Figure 3.1 Distinguishing SOH estimation and forecasting

only enables observing the effect of battery aging. For example, [54–56; 300–302] create models using past and present data. To estimate the SOH they use features derived from the charging curves of current, voltage, and temperature. These curve changes are an effect caused by battery aging. Still, they are hard to interpret and produce for humans. This means a user of the model, like a fleet manager, cannot derive changes of the charging curve (effect of aging) from an assumed usage load of her BEV fleet (cause of aging) with SOH estimation models – even when used for prediction. Only when the effect of battery aging is observable in the future charging curve data, the model can predict the corresponding SOH. Thus, in advance no statement about a future SOH is possible with SOH estimation models.

3.2 State of Health Forecasting vs. Remaining Useful Life Prediction

As introduced in Section 2.1.3, apart from SOH another measure of the battery state is the RUL. SOH and RUL are connected as depicted in Figure 3.2: The SOH is describing the current battery state referring to the initial, new state at the BOL. Contrarily, RUL describes the current battery state by the number of cycles until EOL when a specific state is reached. However, depending on the future load scenario applied to the battery the EOL is reached at a different point in time, as indicated by the dashed blue alternative scenarios in Figure 3.2.

Thus, in contrast to SOH estimation, for a RUL prediction, assumptions on the future load scenario are necessary. Two possibilities for these assumptions exist:

1. First, the same load as in the past is assumed for the future like Gasper et al. [210] do. In this case, SOH estimation has the same information as RUL prediction because the point in time when the EOL is reached is not variable for a single battery. Only the reference point for the state description and the unit are different: SOH in % refers to the BOL, while RUL in cycles refers to the EOL.
2. Second, a different load is assumed in the future than in the past so that the RUL varies depending on the future load scenario shown by the dashed blue lines in Figure 3.2. Thus, additional information about the future load is required by the RUL prediction model, contrarily to SOH estimation models. Under such assumptions SOH and RUL represent different information. This is especially important for the transition form 1st to 2nd life which most likely have different operation conditions. For example, grid applications as 2nd life application are typically more favorable regarding battery aging regarding their thermal mana-

gement systems, reduced SOC windows, and and C-rates compared to EVs as 1st life application [212].

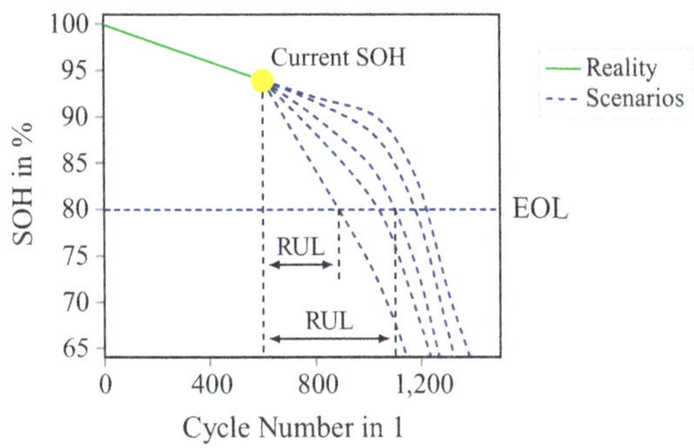

Figure 3.2 Future scenarios of the battery load affect the SOH trajectories and, thus, the RUL

The above mentioned two cases have not explicitly been considered in the literature. The exact opposite, the relation of SOH and RUL is reflected only vaguely: Phrasing like of Tian et al. [303] "there is a strong connection between SOH and RUL" [303, p. 6] suggests that "the SOH and the RUL can be estimated using the same methods [...]." [303, 6]. Considering the two presented cases regarding the future load compared to the past load, this statement holds only true for the 1st case. The 2nd case of RUL prediction is comparable to SOH forecasting since the model requires information about the assumed future load.

Similarly, Ng et al. [304] state that to predict the "SOH [...] spanning many charge/discharge cycles" a function is necessary that "inputs the current state of the battery to predict future behavior" [304, p. 163]. For the prediction of the future SOH in the 2nd case, not only the current SOH has to be input but also information about the future battery load. Such information about the future load can only be ignored if a constant load can be assumed (1st case). Just the same, Hu et al. [305] point out that a lack of missing future measurement results in RUL prediction errors.

In both discussed cases, a RUL prediction model only predicts the number of cycles to a single SOH like EOL_{80}, so multiple RUL prediction models would be required depending on the SOH of interest at EOL. However, a single SOH model can forecast a range of SOHs. From this point of view, SOH forecasting can be seen as a generalization of RUL prediction. This is relevant for BEV fleet operation and also 2nd life applications, as batteries in 2nd life applications usually have a lower SOH (Key criteria 1c & 3 in Table 3.2), for which a RUL prediction model of the 1st life would not be applicable. Additionally, on fleet-level the transition from 1st to 2nd life application is smooth and not necessarily bound to a fixed EOL criterion. This raises the question which output values are appropriate for a RUL prediction model after EOL. Negative values may indicate how many cycles a battery is past the EOL criterion.

Despite, an SOH forecasting model can still be used to obtain the RUL in two procedures: 1. The inputs of the SOH forecasting models contain direct or indirect information on the number of cycles operated. Under this assumption, the inputs can be adapted until, e.g., EOL_{80} is reached similar to the procedure of Lui et al. [306, 5]. 2. Alternatively, the forecasted SOH can be compared with the EOL threshold like 80 %, similar to the procedure of Deng et al. [307, p. 2042]. If the forecasted SOH is greater than the EOL threshold, the RUL is positive. Otherwise, the EOL has already been reached.

Which measure of SOH and RUL to favor, depends on the objective and application of the model's user. If a user has a sense for the quantity of cycles her batteries are cycled per week, RUL may be a suitable measure to temporally classify when the batteries are going to reach the defined EOL. Contrarily, if no specific SOH is defined as EOL threshold, the SOH may be the favorable measure as it can be linked to technical specifications like a minimum required capacity for a certain battery application or a minimum required range of a BEV. Additionally, the RV of a BEV's battery is linked to the offered range, i.e., capacity [308]. Thus, SOH is favorable to indicate RV as well. Additionally, for the transition from 1st to 2nd life application of the battery the SOH as continuous measure is more suitable.

3.3 Scenarios in State of Health Forecasting

Battery SOH forecasting makes statements about the future SOH which is flawed by uncertainty. This uncertainty may have several origins as [309, p. 92] summarizes: Already from production there is intrinsic variability causing variations of the initial state of the battery and also different aging under the same load [310; 311]. When setting up an SOH forecasting model also uncertainty can be distinguished as aleatoric uncertainty inherent to the problem and epistemic uncertainty resulting from limited knowledge or data of the problem [312].[3] An example for aleatoric uncertainty is measurement noise.

[3] For further information consult [312–315].

In this thesis, particularly the uncertainty regarding the future battery load is of interest, i.e., it is unknown how a BEV or BEV fleet will exactly be used. Saxena et al. [316], describe this as operational environment uncertainty which is also identified as a challenge when comparing laboratory to real-world operation by [212, p. 1947]. To address this uncertainty, Greenbank [309, p. 126] proposes to predict the future battery which would require another model. However, this work aims at addressing the uncertainty about the future battery usage by using scenarios to understand the implications of battery usage on battery aging [317]. Scenarios are already common in strategic decision making under uncertainty [317–319]. Porter [317] defines a scenario as "an internally consistent view of what the future might turn out to be."[317, p. 446] Thus, it is not a forecast, but only one possible reality of the future. As Porter [317] notes the assumptions characterizing a scenario might affect and violate each other. The latter can cause the elimination of scenarios.

Parallel to this definition, a usage scenario of a machine is a consistent view of the future usage of a machine. The definition can also be applied to fleets of machines following a definition of the term fleet [6]. Each usage scenario is based on plausible assumptions about important uncertainties regarding the machine's future usage. Building on a single scenario, a set of usage scenarios should reflect a range of possible future usages with important implications for machine aging. Ensuring consistency of assumptions about the future usage within a scenario can be difficult. Especially, physical consistency may be difficult to assess, e.g., the range of a BEV is reduced due to the future load, but the assumption on range reduction may be inaccurate which influences the usage and aging.

Such scenarios can be based on personas or concrete users. Personas give a face to the user by specifying a fictional but still concrete representation of a target user [320] and are a common tool in product development [320; 321]. E.g., a persona may be characterized regarding their battery usage, e.g., by mainly AC- or DC-charging, mainly short trips or long-distance trips on rural roads or highways.

The outcome of scenarios may range from trend to extreme scenarios. Trend scenarios assume the same usage in the future as in the past (ceteris paribus). Positive and negative extreme scenario correspond to the best and worst case respectively. Apart from the outcome of a scenario an estimated probability can express the likelihood of the scenario, i.e., the prediction interval [322].

The time period of the usage scenarios should cover a time horizon in which measurable and relevant aging of the machine can happen. This makes the scenarios distinguishable regarding different aging for the model users. Therefore, for most machines and also for vehicles, this excludes scenarios with a length of seconds or minutes, but scenarios should rather cover weeks, months, or years.

3.4 Battery Model Transfer

Once SOH forecasting gets established as a digital fleet service, the service reliability and availability raises a new challenge concerning the underlying forecasting model: When launching new BEVs or applying the model to another fleet with a new battery type, e.g., with a new cell chemistry, a reliable and usable model suitable for the new battery type is required quickly to support established services.[4]

As will be shown in Chapter 4, existing reliable models for SOH forecasting are data-intensive. However, this is not justifiable for the application in changing environments in most cases, as new data to build a reliable model using these methods cannot be collected in a temporal manner. Hence, due to the aforementioned needed availability their wide acceptance is currently limited. This is especially problematic for new battery types whose data availability is limited in the initial phase: 1. Only few laboratory battery aging test are conducted which suffer under limited comparability to real-world operational conditions (L1 and Section 2.1.6.2) [2; 145; 212]. 2. Usually battery cells, but not packs, systems, or modules are tested in the laboratory. 3. Potentially only a few prototypes or endurance tests may have been operated using the new cell type.

Possible solutions are: 1. To extend this limited data base by further extensive data generation either in the laboratory or using endurance test vehicles. Nevertheless, this solution is expensive and difficult as battery aging is a lengthy process [73, p. 27]. 2. Data synthesis and augmentation from physical aging models still need validation from laboratory tests. Furthermore, an adaption of feature values of aging data and knowing the resulting aging are too complex. The complexity of this causal relationship is the main reason why data-driven models are considered for SOH forecasting. 3. Another solution is reusing physical-chemical models, but no uniform adaption process exists for them which may make adaption complicated and time consuming. 4. Lastly, an established data-driven model of another battery type may be transferred to the new battery type, as soon as a small amount of data of the new battery type has been gathered [6; 275]. Such methods, summarized under the term transfer learning, have been successfully applied in different domains, especially in computer vision [323] and natural language processing (NLP) [324]. Transferability of models to new cells is also mentioned as desirable by [325, p. 7; 326; 309, p. 128], especially for the case without any usage data [309, 129] and for different battery load than of the training data [309, p. 127].

---

[4] This motivation is extended to BEV's apart from the battery model transfer in Appendix G.2.

The application of transfer learning for battery SOH forecasting models is a crucial part, as there are differences in batteries, for instance the nominal capacity, the cell cathode material as well as the applied load due to usage, i.e., the driving and charging behavior in case of BEVs. Over the air (OTA) software updates may affect battery operation, e.g., by increasing the maximum charging power and adaptions of SOC and SOH estimation by the BMS. Even the same battery cell may face different operation when build in different vehicles and powering different electrical machines. However, the general electrochemical behavior of LIBs is a major common characteristic which supposedly provides an excellent starting point for transfer learning. As stated in a previous publication [2], until now transfer learning has not yet been applied to battery SOH forecasting. Because of its importance transferability is required for SOH forecasting models as key criterion 2.

3.5 Model Key Criteria

Now that the batteries and the formal concept of SOH forecasting have been introduced in Section 2.1, 3.1 and 3.2, inspired by the applicability from a user's perspective of BEV fleet managers and battery designers, key criteria that a model for SOH forecasting shall fulfill are derived. These key criteria are from previous publications [1; 2] and summarized in Table 3.2. They can be seen as requirements of SOH forecasting models.

Key criterion 1a:
Assuming a certain load scenario of a BEV fleet, the fleet manager is interested in the future SOH of the fleet's vehicles. The usage load can vary for example due to

Table 3.2 Key criteria for an SOH forecasting model.

Key criterion		
No.	Description	
1a	Contain information about the future load	Model inputs
1b	Information about future load is aggregated	
1c	Capture the higher variability of real-world battery operation compared to laboratory operation	
2	Transfer to new batteries	
3	Applicable for 2nd life applications of batteries	

the charging strategy. This can include the preference for AC or DC charging, i.e., using either the vehicle's on-board inverter or the AC-to-DC rectifier of the charging station before reaching the vehicle's battery, bypassing any on-board inverter. For the fleet manager, it will not be valid to assume the same load in the future as in the past (L2). Thus, models shall have explicit information about the future load as model input (key criterion 1a).

This is coherent with the discussions in Sections 3.1 and 3.2 and the literature: Lucu et al. [285] note that batteries in real-world operation, e.g., in automotive applications, exhibit varying loads during the battery's lifetime. Furthermore, [327, p. 6268; 328, p. 842; 210; 329, p. 218] assume that future battery degradation is independent of the battery's past history. Therefore, information about the future operating load of the battery is essential for SOH forecasting. Fulfilling this key criterion shall enable a model to output future SOH fade or the future SOH based on the current SOH and future load like Lin et al. [330] point out similarly.

Key criterion 1b:
When applying the SOH forecasting model in production, explicit information about the future load is required (key criterion 1a). The future battery load is determined by multidimensional time series of temperature, current, and SOC. For example the United States Advanced Battery Consortium (USABC) sets the development objective for battery life of 15 years and 1000 cycles [33] as introduced in Section 1.1.

This means in real-world operation batteries age slowly, so the relevant time series contain weeks or months of operational data. Furthermore, the sampling rate should be in the range of milliseconds to capture relevant operational dynamics, e.g., different acceleration profiles in city traffic [1]. This temporal scale complicates satisfying key criterion 1a, raising three problems:

1. For forecasting into the future no data from the future battery load is available in the present. Thus, a human needs to express an assumption on the future load via the model's input [75]. Such assumptions can for example regard the charging policy, preferred rest SOC, and limitations of the SOC operating range. Though, the data of the future battery load is difficult to interpret and produce by a human due to its length and multidimensionality [331].
2. In vehicle fleet operation cost reduction is critical for profitability. This also includes costs due to logging operational data of the vehicle fleet as well as transmitting and storing the data, e.g., in a cloud [332, p. 5; 145; 1; 333]. This is a central challenge when forecasting the SOH. Contrarily to these cost factors, computational cost for SOH forecasting is likely to be negligible compared to

running edge computing in the BEVs, if it is provided as a data-driven fleet service in a cloud system [6].

3. The input data of an SOH forecasting model needs to be compatible with the underlying method, e.g., an ML model. The described time series are too long and finely sampled so that, e.g., an RNN would suffer from vanishing gradients [252; 253].

These three problems regarding the data of the future battery load can be solve by data aggregation and feature extraction (Key criterion 1b). This is a common procedure with long time series in general, e.g., in the domain of machine health forecasting [334–337]. Further, it enables humans to run scenario-based what-if-simulation with the SOH forecasting model.

Existing approaches like the one from Song et al. [301] for SOH estimation require data of future charging cycles for prediction like the charging curve of current, voltage, and temperature. These time series change when the battery ages but are hard to interpret for humans. This means a user of the model, like a fleet manager, cannot derive changes of the charging curve from an imagined usage scenario of her BEV fleet. The same applies to the usage use of the incremental capacity (IC) and differential voltage (DV) curves as forecasting inputs [338]. These features are not suitable for SOH forecasting as argued in Section 3.1.

Key criterion 1c:
BEV fleet managers and battery designers are not only interested in battery aging in laboratory experiments, but especially keen on forecasting battery aging in real-world applications. However, as described in Section 2.1.6, the common load of batteries in laboratory and real-world operation differs, e.g., regarding charging and discharging current, temperature, and ΔSOC (L1). Overall, data from real-world operation is more complex and variable than laboratory operation. Thus, the model's inputs about the future load shall be able to capture the higher variability of data from real-world battery operation compared to laboratory operation (Key criterion 1c). This is also pointed out by several authors [75; 180; 213; 217; 218; 236; 240; 305; 325; 331; 339–343].

Key criterion 2:
As motivated in Section 3.4 and stated as major limitation L4 of SotA work, it is beneficial to have the capability to transfer an established model for SOH forecasting of another battery type to a new battery type, as soon as a small amount of data of the new battery type has been gathered [6; 275]. Thus, the required amount of data shall be minimal at least once an initial model has been trained so that it can be

transferred easily to new batteries (Key criterion 2). Transferability of models to new cells is also mentioned as desirable by [325, p. 7; 326; 343].

Key criterion 3:
Once batteries in 1st life applications are not satisfactory anymore, e.g., regarding capacity or the performance desired by the user, they shall be reused in 2nd life applications. Such 2nd life applications like stationary energy storage for grid stabilization usually have less demanding performance key criteria so that older batteries are still suitable. For example, grid applications have better thermal management systems, reduced SOC windows, and c-rates compared to EVs as 1st life application [212]. In automotive applications an SOH of 80 % is often mentioned as threshold for the transition to the 2nd life application [174; 334]. However, manufacturers, fleet managers, and users may choose a different EOL criterion, as pointed out in Section 2.1.3. Also, from a data point of view separate models are debatable: Separating data into two models that include the same aging mechanisms such as lithium plating that become more dominant towards the end of the 1st life would be a drawback for both models. Hence, the model shall be continuously applicable to the 2nd life application of batteries as well, i.e., below an SOH of 80 % [218] (Key criterion 3). This means the SOH forecasting models shall be capable of covering and learning the knee point, if it exists for the given battery type.

When concerned about vehicle replacement, BEV fleet managers are less interested in the SOH forecast after a single cycle or a week, but after some weeks or months once the batteries have aged significantly [60, p. 6]. Thus, also the forecast horizon of SOH forecasting models is tracked, but not defined as an additional key criterion.

Most SOH forecasting methods are providing one-step forecasts, only few output multi-step[5] forecasts [345]. If an SOH forecasting model executes a one-step forecast, multi-step forecasts can be realized by recursive stepwise forecasting as proposed by Liu et al. [346, Table III] and de Falco et al. [347]. Nevertheless, this increases the prediction error due to error propagation. Hence, multi-step forecasting models are preferable.

Intrinsic cell-to-cell variability is caused by manufacturing differences, e.g., due to fouling [310; 325; 348]. However, it cannot be influenced by battery operation, but can be seen as a boundary condition for SOH forecasting that needs to be learned.

[5] These are also called one-step-ahead or single step forecast in contrast to k-step-ahead or k-step forecast.

Related Work

4

Given the delimitation of the object of investigation in Chapter 3 and the key criteria required for a suitable SOH forecasting model defined in Section 3.5, building upon these key criteria in this chapter a structured literature survey for SOH forecasting models is conducted. The related work is structured and discussed regarding the limitations of applicability and comparability with existing models. Also, the limitations of transferability in existing literature on battery aging models are shown in Section 4.3. This chapter concludes with an analysis of the existing research gap in Section 4.4.

The structured literature survey and related work that is presented throughout this chapter are primarily based on previous publications [1; 2; 8].

Within many surveys, different methods for SOH estimation and RUL prediction of batteries are examined and compared [57; 73; 99; 111; 115; 124; 303; 349–351]. The applicability of the reviewed methods is limited for SOH forecasting because they lack input information on the future battery operational load. A study by Badey et al. [217] with similar subject matter is twelve years old so it does not reflect the recent advances in digitization that facilitate enhanced maintenance. Also, this study does not define key criteria for a structured comparison of the different methods. A review by Collath et al. [352] published 2022 is examining methods for aging aware operation of BESS, e.g., for peak shaving or vehicle to grid (V2G) applications.

Supplementary Information The online version contains supplementary material available at https://doi.org/10.1007/978-3-658-43188-4_4.

F. von Bülow, *A Data-Driven Fleet Service: State of Health Forecasting of Lithium-Ion Batteries*, AutoUni – Schriftenreihe 170,
https://doi.org/10.1007/978-3-658-43188-4_4

Their focus is on scheduling of stationary batteries but not on battery operation in automotive applications.

4.1 Screening Method

This literature review is conducted following Webster et al. [353]. First, suitable search engines were identified to obtain the paper corpus: Web of Science [354] searches in relevant journals and conferences in the field of energy storage and batteries. Because Web of Science does not include open-source publications, Bielefeld Academic Search Engine (BASE) was used, while duplicates with the Web of Science search were filtered out. Google Scholar [355] does not enable the mass export of bibliography information and the length of the search string is limited to 256 characters [356]. Using Google Scholar [355] 1 070 000 papers were found.[1] A sample analysis showed that over 90 % of the papers found were categorized as irrelevant only based on the title. Thus, Google Scholar was not used because too many irrelevant papers were found and reproducibility was highly limited. Lastly, semantic scholar [357] was neither used, as it does not enable searching with Boolean operators. Thus, Web of Science [354] and BASE [358] were used to find relevant articles.

Related work found via the search engines was identified based on content analysis. Therefore, a search string was composed of keywords and extended by synonyms of these keywords as listed in Table 4.1. This review focuses on LIBs. Still, methods were included that were originally developed for other battery types because they may be applicable and relevant to LIBs as well. As pointed out in Section 3.1, SOH prediction and forecasting are not clearly distinguishable in the literature. Additionally, the terms "predict" and "forecast" are often used synonymously [293]. Thus, both were considered. As explained in Section 3.2, RUL and SOH can provide similar information. Also, as defined in Equation (2.1.14) the SOH is the relative capacity. Thus, these terms were classified as synonyms. No keyword like "method", "model", "equation", "ML model" or the like is chosen because the review was conducted data-oriented and method-neutral, not method-oriented like others surveys [99; 285]. No limit was applied to the year of publication like "only publications published from 2015 to 2022" in other reviews.

[1] Date: 12.03.2023. With the shortened search string defined by Table 4.1 because of the length limitation of the search string to 256 characters: "(battery OR batteries) (ageing OR aging OR degradation OR deterioration OR "capacity fade" OR "capacity loss") (forecast OR forecasting OR prediction OR predict OR predicting OR prognostics) ("state of health" OR capacity OR "Remaining useful lifetime")".

This resulted in the following search string: ("Battery" OR "Batteries") AND ("Ageing" OR "Aging" OR "Degradation" OR "Deterioration" OR "Capacity fade" OR "Capacity loss") AND ("Forecast" OR "Forecasting" OR "Prediction" OR "Predict" OR "Predicting" OR "Prognostics") AND ("State of Health" OR "Capacity" OR "Remaining Useful Lifetime" OR "Remaining Useful Life").[2]

Table 4.1 Keywords and synonyms for the literature review. Synonyms are connected with the keyword by OR-operator

Keyword	Battery	Ageing	Forecast	State of Health
Synonyms	Batteries	Aging	Forecasting	Capacity
		Degradation	Prediction	Remaining useful lifetime
		Deterioration	Predict	Remaining useful life
		Capacity fade	Predicting	
		Capacity loss	Prognostics	

The filtering process of 2254 papers from the initial search is depicted in Figure 4.1. Based on the defined criteria in Table 3.2, from the initial 2254 papers 1159 papers were filtered by title, 450 papers by abstract, and 579 papers by the full text. Many papers filtered out were about either the estimation of SOH, SOC or investigating battery aging mechanisms. Hence, 66 papers remained for which a forward and backward search according to Brocke et al. [359] was executed which resulted in six and respectively three added papers. The extended corpus of 75 papers was split into 52 research publications, 18 relevant review papers,[3] and five perspective papers. The selected 52 research papers were read carefully regarding the key criteria for SOH forecasting models defined in Table 3.2. As listed in Table C.1 (accessible in the Electronic Supplementary Material), out of the 52 research

[2] Web of Science: Searched using topic which includes the fields title, abstract, author keywords, and Keywords Plus. BASE: Searched using title and subject headings. For searching in BASE the search string was adapted by removing all OR-operators and the quotation marks if the keyword or synonym consisted of a single word: (Battery Batteries) AND (Ageing Aging Degradation Deterioration "Capacity fade" "Capacity Loss") AND (Forecast Forecasting Prediction Predict Predicting Prognostics) AND ("State of Health" Capacity "Remaining Useful Lifetime" "Remaining Useful Life").

[3] One of these review papers is the previously published structured literature review on which this chapters is based on [2].

publications 85 % were journal articles while each 8 % were conference papers and published online, e.g., as PhD thesis on university publication servers (3). The most common journals were the Journal of Power Sources (17 %) and the Journal of Energy Storage (14 %). The papers are spanning from 2010 onward, with over 64 % of the papers published from 2020 onward.

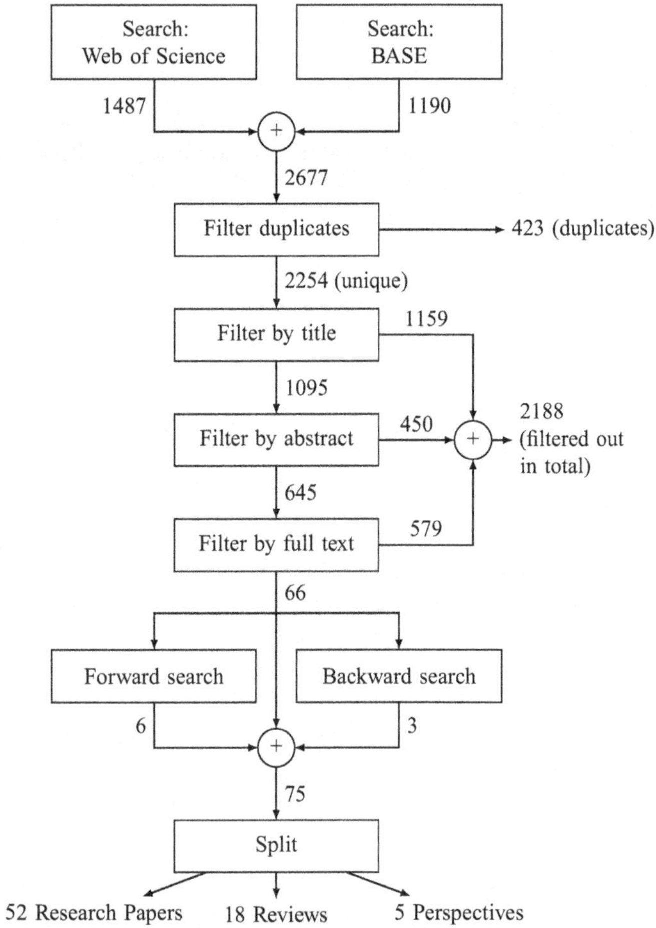

Figure 4.1 Overview of screening method (dated: 06.03.2023)

4.2 State of Health Forecasting Models

When filtering the 2254 initially found papers by title, abstract, and full text, certain categories of models were identified as fundamentally unsuitable regarding the key criteria defined in Table 3.2. Afterwards, the remaining, relevant models were structured and summarized. The unsuitable models are summarized in the following.

There exist many papers [118; 229; 328; 360–369] that present physical, empirical, or semi-empirical models or derive equations for describing battery aging. These models provide a very good indication which operating condition accelerate battery aging. However, these models are based on laboratory experiments executed under controlled conditions in which each ideal cell is linked statically to a specific operation protocol until EOL [370; 371]. In each operational protocol only one or a few operational parameters vary. This limits the model application in real-world operation when multiple operational parameters vary dynamically over several batteries (Key criterion 1) [372; 373]. Karger et al. [168] propose an empirical model with non-global power exponents considering switching operational load after every 100 EFC, but still no explicit input of the operational load is given to their model. Because of the open-loop nature of these models, their generalization and adaptability particularly in long-term predictions is limited [180; 346].

For SOH estimation the equivalent circuit model (ECM) has been discussed by Barré et al. [235]. However, Ng et al. [304] point out that even though computationally efficient, ECMs show limited prediction accuracy across operation conditions in real-life applications that are different to laboratory conditions. Additionally, the difference of SOH estimation and forecasting needs to be considered as described in Section 3.1. In the reviewed literature no clear differentiation of SOH estimation, prediction, and forecast is observed.

Many papers apply particle filters to predict the SOH trajectory, i.e., to predict future battery aging. Less common is the Kalman filter [285]. Both are empirical parametric aging models that recursively update their parameters every time new data is available (update step) and estimate the current state (predict step) [374]. However, estimating a future state would require assumptions on feasible measurement variables. In their review, Lucu et al. [285] list solely models with capacity or internal resistance as measurement variable which are the state variables of interest. So, these approaches are just autoregressive.[4] Autoregressive models output the SOH based on past SOH values [325; 375–381]. This still prohibits SOH forecasting into the far future. Also, no explicit information about the future load is input of these models (Key criterion 1a). Thus, these models assume the same load in the future as

[4] Hu et al. [305] describe autoregressive models as having an "iterative structure".

in the past [309, 5]. However, batteries cycled in real-world operation, for example in automotive applications, may have a varying load over the life of the battery. Furthermore, most autoregressive models only do one-step predictions.

The relevant models found were hierarchically structured into groups as visualized in Figure 4.2 regarding the models' input information of the future battery load. These groups of models are: Those without information about future load often have only the number of cycles k or the charge throughput Q_{thrpt} in ΔAh as input (Section 4.2.1). Those with information about future load have either no distributional information (Section 4.2.2.1) or distributional information (Section 4.2.2.2). Distributional information about future load refers to, e.g., information about the distribution of the charging current. This enables encoding the variability of real-world battery operational data in the model's inputs (Key criterion 1c). Models with distributional information about future load can either include rainflow counting potentially with histograms (Paragraph 4.2.2.2.1) or only histograms (Paragraph 4.2.2.2.2). The models presented in each of the following sections are also summarized in Table 4.3. In each section, first a general introduction is given and then the

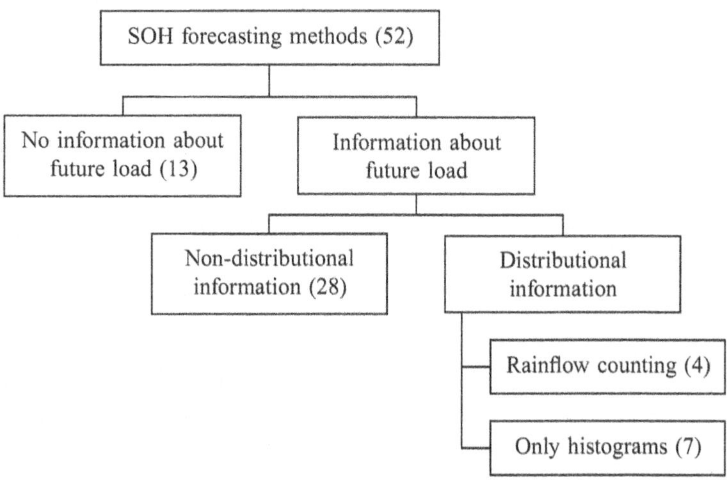

Figure 4.2 Structure of SOH forecasting methods regarding the methods' input information about future battery load. Number of papers in brackets. The 13 papers without information about the future load are representative for this category that includes papers describing physical, empirical, semi-empirical, autoregressive, or equivalent circuit models (ECMs). These methods do not fulfill key criterion 1a because they do not have any information about future load

most cited works from Table 4.3 are presented in more detail. The other works are presented in Appendix C.3 (accessible in the Electronic Supplementary Material).

The majority of the five chosen key criteria is only satisfied by a few methods. All key criteria are only satisfied by [1] with model transferability shown in [8] that will be presented and discussed in Chapters 6 and 7. The models without information about the future load satisfy either key criteria 1a or 1b, but fail at satisfying both. The models with non-distributional information satisfy key criteria 1a and 1b, but fail at key criterion 1c. Only the models with distributional information satisfy key criteria 1a and 1b. Transferability of SOH forecasting has not yet been shown with Gaussian process regression (GPR), but with MLP based on their layer-wise architecture [8] (Key criterion 2). Given the popularity of GPR applied to SOH forecasting, transfer learning with GPR in this field needs further research. If the RUL or cycle life is not used, but SOH or capacity is used as output value, the model is applicable to 2nd life applications (Key criterion 3). So this key criterion is easy to satisfy.

In the context of this review, no uniform symbology has been identified in the paper corpus. To increase comparability, readability, and comprehensibility across paper boundaries, the uniform symbology listed in Table 4.2 is introduced and used.

In Table 4.3, also the open-source availability of code and data is listed which both increase the understanding of the presented methods and allow an easier comparison to other models and other data. As overfitting is a problem in the domain of ML and modeling, it was checked whether authors dealt with overfitting, e.g., by data splitting or cross-validation [382]. The fulfillment of key criteria 1a, 1b, and 1c is dependent on the model's inputs. The vast majority of the found models is applied to LIBs, other batteries types like lead-acid batteries [383] are not so much in the focus of research regarding SOH forecasting.

As observable in the Table 4.3, only a few public data sets were used in the reviewed papers. In more than half of the papers own data was generated and not published afterwards (56 %), which means that the comparability and verifiability as well as the possibility for further use of the results are not given. As Reis et al. [384] and Naaz et al. [385] state, this is a common problem in the domain of battery models. Also, the code is only rarely shared (12 %). Furthermore, data sets of HV LIB packs in BEVs operated in real-world vehicle operation are rare (L1 & L4), coherent to the observation of [386]. Most of the methods use data from LIB cells operated in the laboratory (88 %) confirming the observation of [72]. The few works using data from battery systems or battery packs are listed in Table C.2 (accessible in the Electronic Supplementary Material). A broad range of metrics to measure the models' performance is used. The most used are root mean squared error (RMSE, 39 %), mean absolute error (MAE, 13 %), mean absolute percentage error (MAPE, 13 %), and mean squared error (MSE, 6 %). The most frequently used

Table 4.2 Unified symbology of features identified in the related work

Symbol	Explanation	Symbol used in other papers
k	Number of cycles	n
t_1 and t_2	Beginning and end of time period	
$\Delta t = t_2 - t_1$	Time span from t_1 to t_2	d
C	Capacity	Q, q_{max}
ΔC	Capacity loss	ΔQ
R_{i}	Internal resistance	IR, R, R_0
SOH	State of health or relative capacity	SoH, RC
ΔSOH	Change/loss of SOH or relative capacity	RCL
SOC	State of charge	SoC
$\overline{SOC}$	Middle SOC or average SOC	$\text{Ø}\, SOC$
$\Delta SOC = SOC_{\text{min}} - SOC_{\text{max}}$	SOC lift or SOC-swing	$SOC\text{-}S$
$SOC\text{-}SR = [SOC(t_1), SOC(t_2)]$	SOC swing rate	
$DOD = 1 - SOC$	Depth of discharge	DoD
Q_{thrpt}	Charge throughput in ΔAh	
T	Temperature	
T_{rest}	Rest, hold, or storage temperature	T_{hold}, T_{storage}
T_{ambient}	Ambient temperature	
I	C-rate or current	
U	Voltage	
P	Power	

output values are C (35 %), ΔC (31 %), and SOH (15 %). RUL is only used in 13 % of the identified models.

Additionally, the forecast horizon of the models is diverse and often difficult to interpret and not explicitly given in the reviewed papers. It is measured differently, e.g., in single cycles, multiple cycles, days or as Q_{thrpt} in Ah. In most papers, the forecast horizon seems to depend on the interval of SOH measurements in the used data set.

Table 4.3 Summary of SOH forecasting models. Symbols are defined in Table 4.2. "x" means yes. "-" means no

Ref.	Citations*	Model type	Model inputs $\mathbf{x}^{\dagger}$	Key criterion					Open source		Model performance	Overfitting √	Output y	Forecast
				1a	1b	1c	2	3	Data‡	Code				
SOH forecasting models without information about future load (Section 4.2.1)														
[329]	408	GPR	k	-	-	-	-	x	[71]	-	$RMSE$ = 0.08§	-	SOH	-
[388]	115	Empirical	k	-	-	-	-	x	[71] & -	-	$RMSE$ = 2.5 %	x	SOH	-
[239]	22	Linear	k	-	-	-	-	x	-	-	MSE = 1.7	-	SOH	-
[389]	129	Bayesian	k	-	-	-	-	x	-	-	MAD = 0.02	-	ΔC	1 cycle
[404]	3	LR, RFR, RNN, LSTM	t in d	-	-	-	x	x	-	-	MSE = 3.02	-	$SOH(t_2)$	100 d to 400 d
[405]	31	two vMLPs	$Q_{\mathrm{thrpt,disch}}$	-	x	-	-	x	[68; 71]	-	Not given	-	C and R_{i}	-
[307]	21	Empirical	k, accumulated t_{rest}, k after a regeneration due to t_{rest}	-	x	-	-	x	[71]	-	$RMSE$ = 0.0189	-	$C(t)$	-
[406]	93	Dilated CNN	V_{terminal}, I, T_{cell} of last four cycles	-	-	-/x	-	-	[70]	-	$MAPE$ = 10.6	x	RUL	-
[70]	980	Elastic net	Features derived from the 1st 100 V_{disch} curves	-	-	-	-	-	[70]	x	$RMSE$ = 51 cycles	x	cycle life	-
[325]	64	LSTM	$C(t)$ up to present	autoregressive			-	x	[211]	x	$MAPE$ = 1.1 %	x	$C(t)$	-
[381]	3	RNN	$C(t)$ & $R_{\mathrm{i}}(t)$ up to present	autoregressive			-	x	[70; 211]	x	$MAPE$ = 2.37 % & $MAPE$ = 1.24 %	x	$C(t)$ & $R_{\mathrm{i}}(t)$	-
[407]	15	I.a. CNN	Features derived from the 1st 100 V_{disch} curves	-	-	-	-	-	[70]	x	$RMSE$ = 17 cycles	x	cycle life	-
[408]	1	vMLP	E	-	-	-	-	x	[68]	-	Not given	-	C	-
SOH forecasting models without distributional information about future load (Section 4.2.2.1)														
[177]	43	GPR	Δt, T^{-1}, SOC	x	x	-	-	x	-	-	MAE = 0.31 %	x	ΔC	30 d to 60 d
[390]	54	GPR	Q_{thrpt}, T^{-1}, DOD, $\overline{SOC}$, C-rate$_{\mathrm{ch}}$ & C-rate$_{\mathrm{disch}}$	x	x	-	-	x	-	-	MAE = 0.58 %	x	ΔC	4 kAh to 12 kAh

(Continued)

Table 4.3 (Continued)

Ref.	Citations*	Model type	Model inputs $\mathbf{x}^{\dagger}$	Key criterion 1a	1b	1c	2	3	Open source Data‡	Code	Model performance	Overfitting ✓	Output y	Forecast
[410]	7	SBM	k, I_{disch}, ΔSOC, $\overline{SOC\text{-}SR}$[¶]	x	x	-	-	x	[411]	-	Not given	-	$\frac{dC(t_2)}{dk}$	1 cycle
[346]	197	GPR	$C(t-i), \ldots, C(t)$ with $i = 0, \ldots, 4$, $T_{ambient}$, DOD	x	x	-	-	x	-	-	$RMSE$ = 0.09	x	$C(t_2)$	1 cycle
[327]	8	BNN	C-rate, SOC, T, t	x	x	-	-	x	-	-	MAE ≅ 0.5 %	-	ΔC	-
[412]	19	LSTM	$C(t)$, Δt, SOC_{rest}, T_{rest}	x	-	-	x	x	-	-	$RMSE$ = 0.0008	x	$C(t+\Delta t)$	Δt = 720 h
[393]	63	-	Count events: 1) high I_{ch}, 2) I_{ch} & low T, 3) high I_{ch} & high DOD	x	-/x	-	-	-	-	-	Not given	-	-	-
[413]	9	MFP	$C(k)$, from discharging: $\overline{T_{min}}$, $\overline{T_{max}}$, $\overline{I}$, Δt, λ_{short}, Δt_{rest}, $\overline{V_0}$, $\overline{\Delta V}$, $C_{approx.}$	x	x	-	-	x	[68]	-	$RMSE_{max}$ = 0.1383[‖]	x	$\Delta C(k+m)$	Approx. $m = 50$ cycles
[414]	18	GPR, SVM, LR	k, I_{ch}, I_{disch}, T_{ch}, T_{disch}, T_{rest}, SOC_{rest}	x	-	-	-	x	-	-	$RMSE$ = 0.2956	x	$C(k)$	-
[230]	13	Empirical	$C(t)$, I_{ch}, I_{disch}, T,	x	-	-	-	x	-	-	Not given	x	ΔC	1 s
[424]	10	Hybrid	$\overline{T}$, $\overline{SOC}$, $\overline{I}$, t, F_{rest}	x	x	-	-	-	[71]	-	$\overline{MAPE}$ = 10.15 %	x	RUL	-
[425]	3	TCN	t_{rest}, I, T, V_{cutoff}	x	-	-	-	x	[71; 316; 426; 427]	-	MAE = 0.0295 Ah	x	C	-
[428]	60	Wiener process	$T_{ambient}$	x	-	-	-	-	[316]	-	$RMSE$ = 56.4	-	RUL	-
[340]	5	Wiener process	$T_{ambient}$, C-rate	x	-	-	-	-	- & [429]	-	$\overline{Error}$ = −3.05	-	RUL	-
SOH forecasting models with information about future load based on rainflow counting (Paragraph 4.2.2.2.1)														
[145]	568	SVR	$C(t_1)$, rainflow counting SOC, load collective I over T and SOC over T	x	x	x	-	x	-	-	MSE ≅ 0.0004	x	$C(t_2)$	≅ 8 d to 15 d
[397]	2*	Eq. with four parameters	ΔSOC, $\overline{SOC}$, C-rate$_{ch}$, C-rate$_{disch}$, Q_{thrpt}	x	x	x	-	x	-	-	AE_{max} = 8 %	x	ΔC	≅ 2 kAh to 5 kAh

(Continued)

Table 4.3 (Continued)

Ref.	Citations*	Model type	Model inputs $\mathbf{x}^{\dagger}$	Key criterion 1a	1b	1c	2	3	Open source Data‡	Code	Model performance	Overfitting ✓	Output y	Forecast
[424]	10	Hybrid	$\overline{T}, \overline{SOC}, \overline{I}, t, F_{\text{rest}}$	x	x	-	-	-	[71]	-	$\overline{MAPE}$ = 10.15 %	x	RUL	-
[425]	3	TCN	$t_{\text{rest}}, I, T, V_{\text{cutoff}}$	x	-	-	-	x	[71; 316; 426; 427]	-	MAE = 0.0295 Ah	x	C	-
[428]	60	Wiener process	T_{ambient}	x	-	-	-	-	[316]	-	$RMSE$ = 56.4	-	RUL	-
[340]	5	Wiener process	T_{ambient}, C-rate	x	-	-	-	-	- & [429]	-	$\overline{Error}$ = −3.05	-	RUL	-
SOH forecasting models with information about future load based on rainflow counting (Paragraph 4.2.2.2.1)														
[145]	568	SVR	$C(t_1)$, rainflow counting SOC, load collective I over T and SOC over T	x	x	x	-	x	-	-	MSE ≅ 0.0004	x	$C(t_2)$	≅ 8 d to 15 d
[397]	2*	Eq. with four parameters	$\Delta SOC, \overline{SOC}$, C-rate$_{\text{ch}}$, C-rate$_{\text{disch}}$, Q_{thrpt}	x	x	x	-	x	-	-	$AE_{\max}$ = 8 %	x	ΔC	≅ 2 kAh to 5 kAh
[332]	25	SVR	"Nested" load collective of SOC-classes, T, V, SOC, C-rate	x	x	x	-	x	-	-	Only graphically given	-	$C(t_2)$	≅ 50 d
[398]	31	Empirical: Power law & linear	Histograms by cycle depths & $\overline{SOC}$, profile of T, t_{rest}	x	x	-	-	x	-	-	$MAPE$ = 10.24 %	-	ΔC	1 week
SOH forecasting models with information about future load only based on histograms (Paragraph 4.2.2.2.2)														
[392]	118	GPR	$C(t_1), \Delta t, Q_{\text{thrpt,ch}}, t_1$, T & I histogram-like	x	x	-	-	x	[71]	-	$RMSE$ = 0.0201	x	ΔC	1 cycle
[1]	7	MLP	$SOH(t_1), T$, SOC & I histogram-like	x	x	x	x$^{\Delta}$	x	[70]	-	$RMSE_{\text{test}}$ = 0.084♣	x	ΔSOH	25 to 530 cycles
[401]	44	RSF	5 non-histogram variables, 12 bin variables, 13 derived histogram variables	x	x	x	-	-	-	-	Not given	-	RUL	-
[402]	17	SVR, RFR, GPR, ANN	Statistical properties of histograms of T, $I_{\text{ch}}, I_{\text{disch}}, V_{\text{ch}}, V_{\text{disch}}$, & total Q_{thrpt}	x	x	x	-	x	[68–70] & -	-	$MAPE_{\max}$ = 4.21 %	x	ΔC	1 cycle

(Continued)

Table 4.3 (Continued)

Ref.	Citations*	Model type	Model inputs $\mathbf{x}$†	Key criterion 1a	1b	1c	2	3	Open source Data‡	Code	Model performance	Overfitting ✓	Output y	Forecast
[150]	28	GPR	Histograms of V, T, I, $\lvert I\rvert$, P, $\lvert P\rvert$	x	x	x	-	x	[69; 70]	-	$\overline{RMSE}$ = 0.0017	x	ΔC	Δt = 12 h
[403]	3	Bayesian piecewise LR & GPR	Same as [150]	x	x	x	-	x	[69; 70]	x	$\overline{RMSE}$ = 1.1 % & $\overline{EOL}$ = 1.6 %	x	ΔC & EOL	Δt = 12 h

* From Google Scholar accessed 12.03.2023. † See symbols in Table 4.2. ‡ "-" means non-public data set. § For prediction horizon > 20 cycles ¶ The SOC-SR (SOC swing range) is the range in which SOC-S varies. ‖ Best feature set (MFPc) and worst battery (RW12) results in maximum RMSE. ♯ Only graphical and qualitative results, except of this error for profiles at $SOC = 40\,\%$ after 1 year. # At a prediction horizon of 8 weeks until EOL. ⋆ The work of Nuhic et al. [397] is not publicly available. The first author kindly provided the work upon request. △ In follow-up work [8]. ♦ From Table 12 F-W9-2D, variant A $w_s = 25$.

Overall, the comparability of the performance results of the presented models is difficult because of different operational data sets, different output values, different metrics [387], and different forecast horizons used. Furthermore, a lack of open-source availability of both data [343] and code complicates comparisons. Also, no uniform symbology has been identified in the paper corpus which complicates comparability, readability, and comprehensibility.

ML models were the most used (75 %). Due to the variability of the inputs and complexity of the data, the ML models are able to find patterns and rules in the data. Among the ML models, GPR was the most popular (38 %). Overfitting is not checked by all authors (only 50 %). This is problematic because after fitting models need to generalize well on unseen data.

4.2.1 Models without Information about Future Load

Models without information about future load do not have model inputs regarding the operational conditions I, SOC, and T of the battery in the future (Key criterion 1a). They often only have the time t passed, the cycle number[5] k or the amount of charge throughput Q_{thrpt} as inputs. However, Lucu et al. [285] see such simple SOH forecasting models as "debatable for Li-ion battery aging modeling" [285, p. 95]. The models of this category from Table 4.3 with the highest relevance measured by citations are presented in this section.

[5] Hu et al. [305] describe the structure of such models with $C = f(k)$ as "non-iterative".

Richardson et al. [329] apply GPR with the cycle number k as input to output the corresponding SOH. As the authors themselves note, past data up to a given cycle does not have strong influence on distant future cycles because the model assumes no change in the usage of the battery. Furthermore, the influence of calendar aging on battery aging is neglected. The authors note that future work may include DOD, T, and t_{rest} as inputs which are missing in this work.

Tang et al. [388] use MATLAB's nonlinear curve fitting algorithm to fit the relation of the cycle number k and the SOH with a piecewise cubic interpolation function. When using 30 % of the SOH curve for fitting, a prediction RMSE of 2.5 % is reached.

Guo et al. [389] develop a Bayesian approach for capacity fade modeling. However, the number of cycles is the only covariate. Also, they only perform one-step-ahead predictions.

Severson et al. [70] investigate features derived from the discharge voltage curves from the first 100 cycles to predict the cycle life with an elastic net. Their features are too complex to be human-producible and interpretable (Key criterion 1b). Furthermore, no information about the future load is input of the model (Key criterion 1a).

4.2.2 Models with Information about Future Load

In this section, models are presented whose inputs contain information about the future battery load (Key criterion 1a). First, models without distributional information are presented in Section 4.2.2.1, followed by models with distributional information as model inputs in Section 4.2.2.2.

4.2.2.1 Models without Distributional Information about Future Load

The models in this section only partially satisfy key criterion 1c because of the variability of operational load in real-world which makes distributional information of the operational load necessary. I.e., giving a constant or the mean discharge current of a time window does not provide information about the distribution of discharge current over time.

Lucu et al. [177] propose a GPR model for calendar aging which outputs the capacity loss ΔC with the rest time Δt_{rest}, the reciprocal of the temperature T^{-1}, and the SOC as inputs, where the last two features correspond to the rest time. The model is only considering calendar aging. Furthermore, the inputs are not suitable to capture the variability of real-world operation (Key criterion 1c) because no distributional information of T^{-1} and SOC are given to the model.

In further work, Lucu et al. [390] dissect a GPR model for cyclic aging which outputs the capacity loss ΔC with Q_{thrpt}, the reciprocal of the temperature T^{-1}, DOD, the Middle-SOC (ØSOC), charging, and discharging C-rate as inputs. All features except Q_{thrpt} are corresponding to the cycled Q_{thrpt} in ΔAh. The model is only considering cyclic aging and also lacking distributional information just like the previous work (Key criterion 1c).

Lucu et al. [371] combine their previous works [117; 390] to continuously learn from real-world EV battery operation data, starting with laboratory data. With real-world EV battery operation data the model's predictions improve. As the authors note their model cannot capture the "dynamic character of the DOD stress factor's profiles [with only 20 %, 65 %, and 100 % DOD], which may induce lower aging rates compared to the static DOD aging profiles" [371, p. 5604]. Thus, this model still does not fulfill key criterion 1c.

Andre et al. [391] use support-vector regression (SVR), to learn the relative capacity (RC) which is the $SOH(t_2)$ based on the following inputs: $SOH(t_1)$, T_{min}, T_{max}, time span $\Delta t = t_2 - t_1$, I_{mean}, $Q_{\text{thrpt,ch/disch}}$, SOC_{max}, and ΔSOC. They only state the max. error of 20 % without information on other metrics like the RMSE. When comparing figure 9a) and b) in [391] some overfitting can be assumed. There is no explicit information that overfitting has been tried to avoid.

Liu et al. [346] recognize that Richardson et al. [329] do not consider different T and DOD, while another work of Richardson et al. [392] only did not consider different DOD levels. Thus, based on the later work of Richardson et al. [392], Liu et al. [346] present a GPR model with the current battery capacity $C(t_1)$, optionally the past battery capacity,[6] the ambient temperature T_{ambient}, and the DOD as inputs. Like Richardson et al. [392], their model outputs the capacity one cycle later. For k-multi-step prediction they propose a recursive process of k single step predictions by replacing the model's current and past battery capacity stepwise with the new predicted capacity. However, compared to a single cycle prediction the RMSE of the multi-step prediction increases by 46 % in the best hyperparameter configuration. They do not use information on the SOC, as the $\overline{SOC}$ was 50 %, but only the DOD was varied. Furthermore, no distributional information of T_{ambient} and the DOD is considered.

Rohr et al. [393] suggest three event-counter features which act as stress factors for lithium plating. These features are counting the events of 1) high I_{ch}, 2) high I_{ch} & low T, and 3) high I_{ch} & high DOD. Neither do they specify what "high"

---

[6] $C(t - i), \ldots, C(t)$ with $i = 0, \ldots, 4$.

means for the three signals nor is discharging information included. Their work only proposes features concerning cyclic aging and charging for SOH forecasting but does not apply them on a data set or model.

Safari et al. [394] present empirical third-order polynomials with zero intercept as fitting functions with aging time t, I_{ch}, I_{disch} and SOC as inputs. They fit the model parameters at BOL and after every 3 months for 1 year. Neither temperature information, nor any distributional information is input of their model (Key criteria 1a & 1c).

4.2.2.2 Models with Distributional Information about Future Load

Models with distributional information about the future load can either include rainflow counting (Paragraph 4.2.2.2.1) or only histograms (Paragraph 4.2.2.2.2). Models with rainflow counting may also contain inputs structured as histograms. Histograms in the context of battery aging are also referred to as load collectives and are known from material science and mechanical engineering. Similar concepts are dwell time and residence time counting. Wenzl et al. [75] point out that histograms enable easy analysis of how changes in the operational conditions of a battery affect the aging by adapted, synthetic histograms. This corresponds to the first problem motivating key criterion 1b in Section 3.5.

4.2.2.2.1 Models with Information about Future Load based on Rainflow Counting

Rainflow counting is known from damage evaluation of metals and mechanical fatigue analysis [395]. But it can also be used for SOC cycle counting, i.e., to count how often the SOC jumps from one level range to another [396]. For example, SOC changes from the range of 0 % to 10 % SOC to 10 % to 20 % SOC could be counted. A figurative description is to imagine rain drops flowing and falling down a pagoda roof. Rainflow starting from different levels may interrupt each other creating sub-cycles. Rainflow counting enables the consideration of micro-cycles which has been mentioned in Table 2.3 as a difference of laboratory and real-world battery operation.

Nuhic et al. [145] output the future $C(t_2)$ based on the current $C(t_1)$, rainflow counting of SOC as well as different load collectives. The load collectives are 2D-histograms of I & T and SOC & T. Nuhic et al. [145] use data of six batteries operated in laboratory with real-world driving profiles at different T, C-rates, SOCs, and DODs. Their method learns the degradation behavior of the batteries, but transferability is not given because SVR was used.

Based on their previous work [145], Nuhic et al. [397] introduce an improvement by adding and combining input information regarding the ΔSOC, ØSOC, C-rate$_{ch}$, C-rate$_{disch}$, and Q_{thrpt}. In this work, for simplicity they only consider $\Delta SOC >$

2 % with the rainflow counting method, but smaller ΔSOC are considerable for recuperation of BEVs. Nuhic et al. [397] also identify that a simplification of the current is necessary to prevent a large amount of data. They propose the calculation of a constant, mean C-rate for discharging and charging for each cycle determined with the rainflow counting method. After estimating the four parameters of the defined model equation, a maximum error bound of 5 % is reached. For a modified current profile, the error bound increases to a maximum of 8 %. The authors note that their work is limited to cyclic aging only, but calendar aging should be considered in future work. In contrast to most other work found, this work uses a SOC profile inspired by real-life current profiles in a one-year laboratory test.

Building on their previous work [145; 397], Nuhic et al. [332] add "nested" load collectives. These include nested histograms for SOC, $\overline{T}$, $\overline{V}$, and C-rate based on separate SOC cycles determined by the rainflow counting known from Nuhic et al. [145]. E.g., a nested histogram is structured as follows: First, the cycle's $\overline{T}$ is binned, e.g., from 20 °C to 30 °C. Second, within this bin the cycle's $\overline{V}$ is differentiated into further bins, e.g., from 3.8 V to 4.1 V. Finally, within this bin a 2D histogram with bins of C-rate and SOC is created. So overall a 4D histogram is created. The histograms in their previous work [397] are based on separate SOC cycles determined by the rainflow counting. This means only the time window of a certain SOC cycle is considered for counting in the histograms. This is in contrast to Richardson et al. [392] and the previous work [1], who use the time window of the complete forecast horizon at once for their histograms. A further difference is that Nuhic et al. [332] count the amount of SOC cycles having values like $\overline{T}$ in a certain range, but the histograms of Nuhic et al. [145], Richardson et al. [392], and the previous work [1] count the time spend in an operational state. When considering rainflow counting separately from the histograms like in [145], information on the SOC or voltage level of the (micro-)cycles is lost. This is solved in the last work of Nuhic et al. [332] by using nested histograms which connect the rainflow counting with voltage information.

Stroe [398] proposes a method in his PhD thesis that is based on rainflow-cycle counting. All identified cycles are summarized in a histogram by their cycle depth and $\overline{SOC}$ as well as a separate T profile and rest time t_{rest}. Precise information on I and T is not input of the model so important stressors in real-world operation are missing (Key criterion 1c). Also, the method assumes path independence through the Palmgren-Miner rule [399; 400].

4.2.2.2.2 Models with Information about Future Load only based on Histograms

Richardson et al. [392] present a GPR model with current $C(t_1)$, total time elapsed during the load $\Delta t = t_2 - t_1$, the charge throughput $Q_{\mathrm{thrpt,ch}}$, the absolute time elapsed since BOL t_1, and lastly histogram-like features. These histogram-like features are the time elapsed during which certain conditions regarding intervals of T and I are met. E.g., one input is the time elapsed during 2 A to 3 A. The model's output is ΔC one cycle later so consequently the histograms are also created only for a single cycle. However, the bins of the model's inputs are very broad and do not consider charging information separate from discharging. E.g., the model does not distinguish between different charging currents as these are all considered in the current range above the discharge current of −2 A.[7] Thus, the model requires adaption before being able to capture the variability of real-world data (Key criterion 1c).

Based on this, in a previous publication [1], further presented in Chapter 6, MLP models were trained with $SOH(t_1)$ and combined histograms of T, SOC & I as inputs. The ΔSOH is forecasted over multiple cycles based on histograms of the operational data of these cycles. The histograms from Richardson et al. [392] are one dimensional (1D) histograms, e.g., a bin is $2\,A < I < 3\,A$. These 1D histograms were extended to 2D and 3D histograms which combine bins of two or three signals respectively. For example, a bin is $0C < I < 3\,\mathrm{C}$ and $28\,°\mathrm{C} < T < 31\,°\mathrm{C}$. The examined 2D histograms are T & SOC, I & SOC, and I & T (Variant A) and T & SOC, I & SOC (Variant B, like variant A without I & T). The 3D histogram has T & I & SOC. The authors examined different sampling granularities of the histograms and find that 2D histograms work sufficiently well compared to 3D histograms.

Equivalent to the 2D histogram-like features of [1], Nuhic et al. [145] also propose two parameter instantaneous value (dwell time-) counting combining I & T as well as SOC & T. However, they combine it with rainflow counting so that this work is presented in detail previously in Paragraph 4.2.2.2.1.

Frisk et al. [401] apply random survival forest (RSF) to estimate the RUL of lead-acid batteries. They express the RUL as function $B(t; t_0, V)$ which describes the probability that the battery of a specific vehicle V will last at least t time units after t_0. They use data from 33 603 Scania heavy-duty trucks from 5 European markets with 291 variables including 17 histograms. Of these they select the 30 most important variables using an area under the curve (AUC)-based score. These include among

[7] [392] do not use the sign convention of this thesis and IUPAC. I.e., the unified value of −2 A corresponds to the original 2 A in their work.

others: The total distance driven, the truck's time of delivery, the number of days in use and histogram values like the time spend at low battery voltage at relatively low temperatures of -5 °C to 10 °C, at low battery voltage in general, as well as low ($< 20\,\%$) and high ($> 80\,\%$) SOC without battery load. They point out that it is not sufficient to build a RUL prediction model only on calendar time and mileage. The authors do not provide a performance metric of their final model.

Zhang et al. [402] examine SVR, random forest regression (RFR), GPR, and ANN with statistical properties of histograms as features. The statistical properties of the histograms are: Maximum, minimum, range, average, variance, skewness, kurtosis, 5th moment, and 6th moment. Spearman correlation analysis is used to identify the relevant statistical properties that serve as model inputs. The histograms are T, I_{ch}, I_{disch}, V_{ch}, V_{disch}, and total Q_{thrpt}. They use three public data sets [68-70] and a non-public real-world vehicle data set. For the vehicle data set additionally the SOC is differentiated in separate histograms for charging, driving, and parking mode, T is differentiated for parking and driving mode, and DOD is added. However, the statistical properties of the V-histogram are not used. As further adaption, the accumulated parking time and total time duration are added. RFR outperforms SVR, GPR, and ANN on all data sets. Its maximum MAPE is 4.21 % on the NASA Randomized Battery Usage data set [68] presented in more detail in Section 5.1.3, and 1.41 % on the real-world vehicle data set. Model transfer from one data set to another has not been examined yet (Key criterion 2). A challenge will be that the most correlating features are not the same for different data sets.

Greenbank et al. [150] generate histograms of V, T, I, absolute current $|I|$, power P, and absolute power $|P|$. They determine the size of the bins from the 1st, 33rd, 67th, and 99th percentiles of these histograms and derive their features from these percentiles. Disadvantageously, this can only be done after all signal values have been recorded. So, the advantage of time series aggregation in histograms regarding data saving and transfer is lost. Compared to other histogram-based features like in [1], Greenbank et al. [150] have overlapping bins: For example, the second voltage bin is from 2 V to 3.51 V, which is simply the sum of the values of the first bin from 2 V to 3.12 V and the forth bin from 3.12 V to 3.51 V. Overall, they generate 74 features which they reduce to 5 features using Pearson's correlation coefficient. As inputs of a GPR these features reach a mean RMSE of 0.17 %. The same work appears also in the PhD thesis of Greenbank [309].

Greenbank et al. [403] build upon their previous work [150] using the same features of the time spent within percentile bins. This rises the same problem of determination of the percentile bins as in previous work. Compared to their previous work, in this work they use a data set with a reduced number of features and samples and find that a piecewise-linear model reached a comparable performance to a GPR model on this data set. The piecewise-linear model using Bayesian linear regression (LR) in each segment are meant to be simple and flexible compared to other models.

4.3 Transfer Learning for Battery Models

Transfer learning is well established in the field of computer vision as pointed out in Section 2.2.5. However, for transfer learning applied to SOH forecasting few direct related work exists as observable in Table 4.3 in the column showing the satisfaction of key criterion 2. Still, in the initial paper corpus of 2254 papers only 16 have the term "transfer learning" in their title. Based on these papers a short, unstructured literature survey was conducted which is not following the structured screening method presented in Section 4.1. Also, the topic was extended beyond SOH forecasting to battery aging in general. From the context transfer learning and battery aging three existing works are presented:

Azkue et al. [430] trained a baseline MLP using only calendar aging data of 30 NMC cells. Then they apply transfer learning with a reduced amount of another LFP data set. As benchmark, they trained a model solely on the LFP data set. They achieved a better performance with two LFP cells and transfer learning than without transfer learning and five LFP cells. Moreover, their results indicate improved generalization with transfer learning. However, they only compared to one benchmark model and used only one target data set. In addition, the transfer is only examined for calendar aging, not cyclic aging.

Shen et al. [326] trained a CNN for capacity estimation of LIBs on a source data set of ten years of cyclic aging data. All five convolutional layers and the three fully-connected layers of the pretrained model were transferred. The best performance was achieved, when fine-tuning all layers providing the most capabilities for model adaption to the target domain. Compared to Azkue et al. [430], Shen et al. [326] vary the amount of target data available for fine-tuning. They find that training a benchmark model from scratch needs three times more training samples compared to transfer learning. Thus, transfer learning saves time and costs for data collection in their application.

Tan et al. [431] used features from the voltage curve during CC charging. Their source model is trained only on the battery "B7" of the NASA Battery Data Set [71]. They then transfer their LSTM model to other batteries of the same data set with NMC as cathode material and four batteries of the CALCE data set [316] with lithium cobalt oxide (LCO) as cathode material. So their transfer is once considering the same cathode material and once a different. The prediction was improved when using only the first 25 % of data of a new battery.

These three papers indicate that transfer learning is a promising approach for battery aging models. However, none of these papers considered the temporal sequence when data becomes available for training. This consideration is important to answer the question "when to transfer." Further, each paper only used one target data set. Thus, they cannot be used to compare the influence of different cell chemistry and battery operation on the success of transfer learning. In summary, transfer learning requires more research in the domain of modeling battery aging and SOH forecasting.

4.4 Gaps of Related Work

The analysis of the related work gained insights in possible input values of SOH forecasting models. Also, rain flow counting and histograms as time series aggregation methods maintaining distributional information for SOH forecasting models were presented. Only 11 SOH forecasting models use distributional information about the future battery operational load. The majority either cannot input information about the future load or the information is only non-distributional. The latter means that constant load, e.g., regrading ΔSOC, T_{rest}, and SOC_{rest} is assumed for the whole forecasting horizon. This is an unrealistic assumption in real-world operation of BEVs. In existing literature, only ML models like SVR, GPR, and MLP were shown to work with distributional information of battery operation. In contrast to humans and other models, ML models are able to learn the complex dependencies of battery operation and battery aging from data.

Overall, the use and development of SOH forecasting models in real-world operation according to the defined key criteria are still in an early stage. However, the evaluation of the found methods is only regarding the finite set of the five chosen key criteria in Section 3.5. Other key criteria, e.g., related to the technical implementation of the methods are not considered. One could consider the necessary amount of training data, required computational resources for training and interference, and size of the model (in storage MB or number of parameters). These key criteria are out of scope for this review but will become important once the field of SOH forecasting advances.

As displayed in the Venn diagram in Figure 4.3, the majority of the five chosen key criteria is only satisfied by a few methods. This opens a research gap, especially for the model transfer (Key criterion 2) which has only been focused in the previous publication [8]. But also intelligent aggregation of battery operational data applicable to real-world data needs further research (Key criterion 1b). Especially for changing battery load there is limited research on SOH forecasting

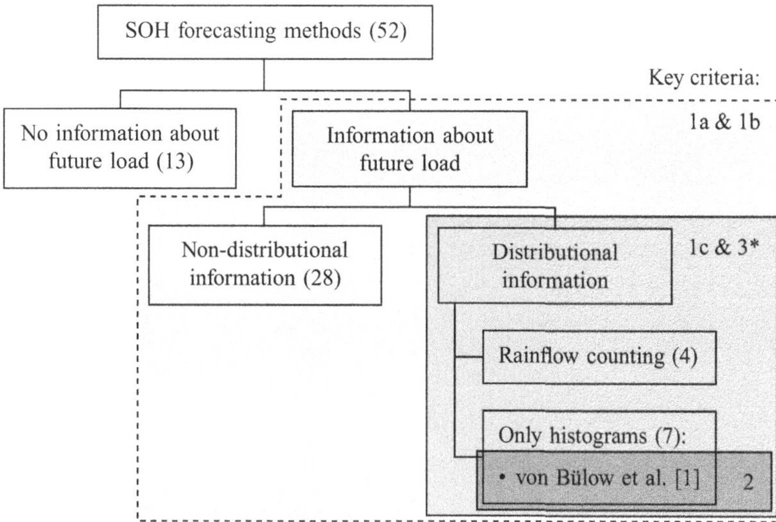

Figure 4.3 Venn diagram based on method hierarchy regarding input data of the methods (see Figure 4.2). The fulfilled critical key criteria correspond to the marked areas. The key criteria are defined in Table 3.2.
*Key criterion 1c and key criterion 2 are only not fulfilled for [392; 398; 401]. Some of the methods with non-distributional information do not fulfill key criterion 1b.

(Key criterion 1c) [240]. Concluding the analysis of the related work, the development of a new approach for SOH forecasting in real-world applications is necessary considering the defined key criteria in Table 3.2.

Data 5

The training data suitable for SOH forecasting should be obtained from several batteries which have aged under a wide range of operational scenarios. This enables the model to forecast the SOH of batteries given an aging scenario encoded in the stressors.

In this thesis, different data sets are used: Five public battery data sets from laboratory operation are presented in Section 5.1. One non-public battery data set from BEV fleet operation is introduced in Section 5.2. These data sets have already been presented in the previous publications [1; 4; 8] and [3; 9] respectively.

Currently, no standard data set from real-world operation exists for battery SOH forecasting models analogous to ImageNet [432], MNIST [433] or CIFAR [434] for image classification models. Such open-source data sets give different researchers the opportunity to compare the performance results, e.g., of their data augmentation techniques [435].

The SOH_{C} and the related SOH_{E} were introduced in Equations (2.1.14) and (2.1.15) respectively. For the laboratory data presented in Section 5.1 and used in Chapters 6 and 7, the SOH_{C} is used given its availability. For the BEV data presented in Section 5.2 and used in Chapter 8, the SOH_{E} is used given its availability. For simplicity both, SOH_{C} and SOH_{E}, are referred to as SOH in the respective chapters.

Supplementary Information The online version contains supplementary material available at https://doi.org/10.1007/978-3-658-43188-4_5.

F. von Bülow, *A Data-Driven Fleet Service: State of Health Forecasting of Lithium-Ion Batteries*, AutoUni – Schriftenreihe 170,
https://doi.org/10.1007/978-3-658-43188-4_5

5.1 Laboratory Battery Cell Data Sets

For each data set in the overview given in Table 5.1, this section first presents a brief description and the SOH trajectories in Figure 5.1. Then, if necessary, applied data cleansing methods are described, if physical inconsistencies in the data like time jumps or measurement errors were found. Lastly, cells that were not used and the reasons for this decision are specified. The data sets presented throughout this section are already presented in the previous publications [1; 4; 8].

The following two procedures were applied to more than one data set:

1. For NASA random and Oxford, the start of a new cycle is not given. Thus, a cycle start is set when the current in a new step switches from zero or below zero (storage or discharging) to higher than zero (charging) so that every cycle starts with charging. As well, the previous cycle needs to be at least a complete EFC before a new cycles starts.
2. For RWTH Aachen ISEA (ISEA), NASA random, and Oxford, capacity measurements are conducted in regular intervals. The SOH values are interpolated over time in between these capacity measurements because the aging processes between capacity measurements are continuous.

5.1.1 Stanford Battery Data Sets

[70] and [69] present public fast charging data sets using the same battery cell types with varying fast charging protocols. Both are referred to by the paper's name first using the data set as Data-Driven and Closed-Loop respectively. These commercial LFP/graphite cells were manufactured by A123 systems (APR18650M1A). They were cycled at a constant ambient temperature of 30 °C under varied fast charging conditions but identical discharging conditions with a C-rate of 4C. The cells have a C_{nom} of 1.1 Ah. The sampling time is 1 s. The SOH values are defined by the discharged electric charge of the corresponding cycle. The SOC signal is calculated offline relatively to the discharged electric charge of the previous cycle.[1] SOH values corresponding to single cycles are identified as outliers when they deviate more than three local SDs from the local mean within a 30-element window of the neighboring SOH values. These values are interpolated linearly using the neighboring SOH values.

[1] The discharged electric charge is specified by the end value of the signal "Qd".

Table 5.1 Overview of laboratory battery data and their relevant characteristics. Cathode and anode materials are explained briefly in Appendix A.1.1. So are the form factors in Appendix A.1.2

Data set	Characteristics	Data-Driven (DD)	Closed-Loop (CL)	RWTH Aachen ISEA (ISEA)	NASA Randomized Battery Usage Data Set (NASA random)	Oxford Path Dependent Battery Degradation Data Set (Oxford)
Source Section		[70] 5.1.1.1	[69] 5.1.1.2	[211] 5.1.2	[68] 5.1.3	[169] 5.1.4
Battery aging	Number of cells	42	45	48	26	27
	Life time range in cycles	532-2,235	530-1,231	1,410-1,872	199-1,188	300-1,681
	Min. SOH in %	80-881	22-761	501	50-861	2-86
Battery type	Nominal capacity in Ah	1.1	1.1	≅1.85	2.1	3
	Nominal voltage in V	3.3	3.3	3.7	?	3.6
	Voltage limits in V	2.0-3.6	2.0-3.6	3.0-4.1	3.2-4.2	2.5-4.2
	Form factor	18650 (cylindrical)				
	Cathode material	LFP	LFP	NMC	LCO	NCA
	Anode material	Graphite	Graphite	Carbon	Graphite	Graphite
Battery operation	Max. charging C-rate	3C to 8C	3.6C to 8C	2.2C	2.2C	0.5C
	Max. discharging C-rate	−4C	−4C	−2.2C	−2.4C	−0.5C
	ΔSOC in %	100	100	60 (20–80)	Up to 100	100
	$T_{ambient}$ in °C	30	30	25	20 & 40	24
	Storage time	0 h	0 h	0 h	0 h	5 d or 10 d

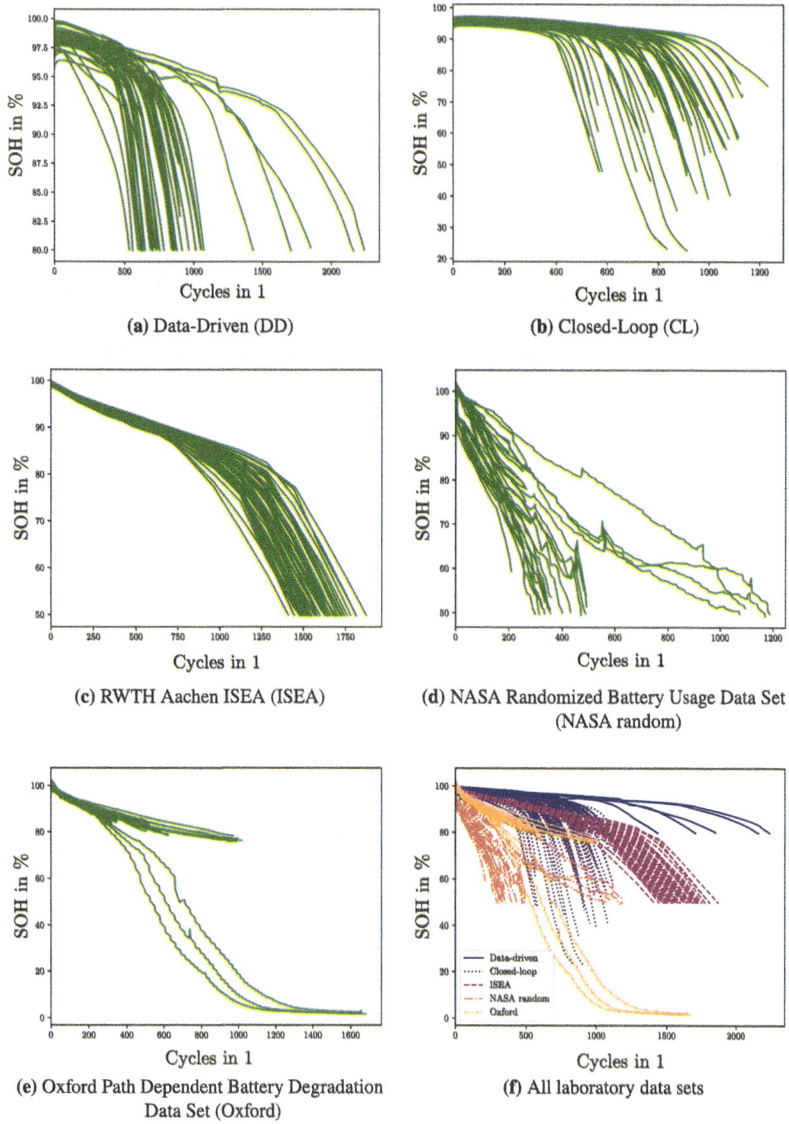

Figure 5.1 SOH over cycle number k of all used laboratory data sets. Discontinuities appear because of plotting over cycles, but the SOH is interpolated over time, not over cycles

5.1.1.1 Data-Driven

The Data-Driven (DD) data set [70] used in this thesis consists of 42 LIB cells belonging to the batch of 2017-05-12 cycled up to 80 % SOH. All cells are charged with a two-step fast-charging protocol. This protocol has the format "C1(Q1)-C2", in which C1 and C2 are the C-rates of the first and second CC steps (CC1 and CC2 respectively) and Q1 is the SOC at which the current switches. C1 and C2 range from 3C to 8C, while Q1 ranges from 15 % to 80 % SOC.[2] The second current step ends at 80 % SOC, after which the cells charge with another CC step at 1 C (CC3) followed by a CV phase.

The SOH signal is smoothed using a moving average with a span of 15 data points. The cells 1 to 5 of batch 2017-05-12 continued as cells 8 to 17 in batch 2017-06-30. Thus, they are concatenated in batch 2017-05-12. The thermocouples of cells 15 and 16 in batch 2017-05-12 were switched. This means that the temperature signals of these cells are not synchronized with the other cells' signals. It was not possible to synchronize them. Thus, these cells are not used. Neither are cells 1 and 19 used as some of their cycles are up to 11.8 times longer as the previous cycles because the switching from charging to discharging happens late. This would skew the min-max-normalization. Cells 3, 7, and 8 are used as validation cells which means that they are not part of the training, validation, or test set.

5.1.1.2 Closed-Loop

The Closed-Loop (CL) data set [69] used in this thesis consists of 45 LIB cells belonging to the batch of 2019-01-24 cycled up to 22.8 %. The cells in the other batches are only operated for 100 to 120 cycles. All cells are charged with a three-step fast-charging protocol. This protocol has the format "CC1-CC2-CC3", in which CC1, CC2, and CC3 are the C-rates of the first, second, and third CC step that end at 20 %, 40 %, and 60 % SOC respectively. CC1, CC2, and CC3 each range from 3.6C to 8C. A fourth parameter, CC4, is dependent on CC1, CC2, CC3, and the charging time. CC4 ranges from 2.68C to 4.755C.[3]

The SOH signal is smoothed using a moving mean with a span of 25 data points because the SOH signal is noisy. Cells 3, 7, and 8 are used as validation cells which means that they are not part of the training, validation or test set.

---

[2] C1 and C2 $\in$ {3, 3.6, 4, 4.4, 4.8, 5.4, 6, 7, 8}C. Q1 $\in$ {15, 25, 30, 35, 40, 50, 60, 70, 80} % SOC.

[3] CC1, CC2, and CC3 $\in$ {3.6, 4.4, 4.8, 5.2, 5.6, 6, 7, 8}C. CC4 $\in$ {2.68, 3, 3.652, 3.834, 3.94, 4.16, 4.252, 4.755}C.

5.1.2 RWTH Aachen ISEA Cyclic Aging Data Set

The ISEA data set [211] consists of 48 LIB cells cycled with the same profile under equal conditions to explore intrinsic cell manufacturing variability and small temperature differences in battery packs during operation. These cylindrical NMC/graphite cells, manufactured by Panasonic/Sanyo (UR18650E), were cycled at a constant ambient temperature of 25 °C. Each cycle starts with 30 min discharging to 3.5 V followed by 30 min charging to 3.9 V. Both currents are limited to a maximum of 4 A. The charge turnover is about 1 Ah corresponding to cycles between approximately 20 % and 80 % SOC. Because of the voltage limits the charge turnover varies with the SOH of the cell, but the DOD relative to the current capacity C is almost constant. These C-graded cells are from the same production lot. Due to this factory selection the cells only have a mean capacity of approximately 1.85 Ah which is used as C_{nom}. Capacity measurements are on average conducted every 165 cycles. The sampling time is on average 0.1 s.

The SOC signal calculation is orientated at the given boundaries of approximately 20 % and 80 % SOC for each cycle. Jumps of the time signal where corrected for cell 3 and 8. If the temperature signal is 0 °C for a complete cycle, the ambient temperature of 25 °C was set because this is the lowest physically possible temperature. This ensures that these states are considered in the stressor tables at least in the ambient temperature bin. If the temperature dropped to 0 °C only for some time stamps, it was interpolated using the neighboring values because sudden temperature breaks are seen as impossible as the temperature only changes slowly. Cell 47 has a noisy temperature signal which was smoothed using a Fourier transform (FT) and moving mean. For some cells the last cycle length is 1820 min instead of 60 min. For cell 29 and 18 the last 3 and 86 cycles respectively are very short. Because of these irregularities, only the data corresponding to SOH higher than 50 % is used.

5.1.3 NASA Randomized Battery Usage Data Set

NASA random [68] consists of 26 LIB cylindrical (18650) cells of type LCO/graphite by LG Chem. that were cycled in seven groups with different cycling protocols for each group (For details see Table D.1, accessible in the Electronic Supplementary Material). The cells are cycled with randomized current sequences ranging from 0.5 A to 4 A. The randomization is meant to approximate real-world battery load. The temperature was environmentally controlled. The sampling time is 1 s. Every fifty cycles the capacity was determined with two measurements.

Some capacity measurements were incomplete, resulting in a lower capacity value. Thus, if the capacity values deviate more than 0.1 Ah from each other, the higher value is chosen. The SOC signal calculation is orientated at the voltage limits after charging and discharging. Voltage jumps from 3.2 V to 4.2 V combined with a time jump in the data are likely due to missing charging data. For best correction, values with the sampling time of 1 s were added to the data set: The current so that the battery is fully loaded with CC. Temperature and voltage are linearly interpolated. Else the average of both is used. Batteries RW2 and RW18 were not used because the temperature signal has many values of −4000 °C which lies below absolute zero.

5.1.4 Oxford Degradation Data Set

Oxford Data set [169] consists of 27 LIB cells of type Panasonic lithium nickel cobalt aluminum oxide (NCA)/graphite 18650 that were cycled in four groups with different cycling protocols for each group.[4] All cells were cycled with CC between 0 % and 100 % SOC. After a period of cycling, calendar aging was performed at 90 % SOC. The time ratio of cyclic to calendar aging was 1:5 with different C-rates and storage SOC (Details in Table D.2, accessible in the Electronic Supplementary Material). Compared to the other data sets, only the Oxford data set contains calendar aging. The temperature was environmentally controlled at 24 °C. Capacity measurements were conducted every 48 cycles by CC-CV charging until $I = C/60$. The sampling time is 1 s.

The SOC signal calculation is orientated at the voltage limits after charging and discharging. Again, temperatures below 0 °C and Not a Number (NaN) values are replaced by the ambient temperature of 24 °C to minimize adulteration of the training data ($T_{\min} = 24$ °C, $T_{\max} = 36.6$ °C, $\overline{T} = 25.2$ °C). Inconsistency like negative time deltas and time jumps of "TestTime" higher than the sampling time with following constant time values were corrected by replacing the time signal with the sampling time of 1 s onwards. This sampling time was determined from the consistent time signals. This adaption seems legitimate, especially for the time jumps, because the duration of the time jumps equals the product of sampling time and the amount of following constant time values. The file no. 27 of battery number 9 from part 2, group 1 has some single current values around -10^{282} A during CC phases. These were replaced by the average of the two neighbor current values. Battery no. 1 from part 2, group 6 was not used because it is only exposed to calendar aging which prohibits the identification of cycles. Battery no. 10 from part 1, group 2 was not used because it only contains a single cycle. The file no. 14 and 15 of

[4] The three parts of the data set can be found here: [436–438].

battery no. 14 from part 1, group 3 are renamed and used for battery no. 15 of the same part and group because these are the only files of battery no. 14. At the same time exactly these file numbers are missing for battery no. 15.

5.2 Real-World Battery Electric Vehicle System Data Set

In this section SOH means SOH_E and not like in the previous sections of this chapter SOH_C as defined at the beginning of this chapter.

Using battery system operational data from real-world BEV operation, i.e., with NI current profiles over multiple cycles and VC (NI-VC) as introduced in Section 2.1.6.2, seems to become feasible as the registrations of BEVs have increased worldwide in the past years [19]. But the majority of these BEVs are used privately and are parked most of the time, e.g., approximately 89 % to 97 % in Germany [221; 439; 440]. Battery aging is correspondingly slow, so data from BEVs operated in such a way will only be suitable for creating SOH forecasting models in the future. In contrast to the domain of private BEVs, BEVs in MOD use cases are operated more intensively and age quicker [2; 225]. Therefore, data from the domain of MOD BEV fleets is more suitable nowadays to apply SOH forecasting on real-world BEV operational data.

The data set originates from battery systems of 550 BEVs operated in a ride-pooling MOD shuttle service in two large German cities. The total time span of data logging is 27 months including a total of 7 months operational pauses due to COVID-19 restrictions. During the total time span the vehicles drove a total of $39.1 \cdot 10^6$ km corresponding to 115 239 EFC. Based on the average total ΔSOH and the total EFC, the mean SOH decline is 0.017 % SOH per EFC. The energy throughput by charging was 53 % AC and 47 % DC with up to 7.2 kW and 100 kW charging power respectively. I.e., the highest charging C-rate of 1.25 is comparatively moderate.

The HV system of the BEV is based on NMC-622 pouch cells. The BEV has a nominal total energy content $E_{nom,pack}$ of 87 kWh and a nominal capacity $C_{nom,pack}$ of 240 Ah. The BEV enables storing braking energy (braking recuperation) as well as coasting energy (coasting recuperation) of the motors. Via a DC/DC converter, the HV battery system also supplies the 12 V on-board electrical system for air conditioning, USB chargers, etc.

The sampling time and signal resolution of the relevant time series signals are given in Table 5.2.[5] The BEV-mode is driving (discharging or charging, i.e., recu-

[5] A introduction into automotive time series data is given in Appendix B.2 (accessible in the Electronic Supplementary Material).

Table 5.2 Details of used signals

Signal	Sampling time in s	Resolution	Calculated offline
Current I	10	1 A	–
Temperature T	$\approx$ 120	0.5 °C	–
SOE	10	0.5 %	–
Energy E	10	50 kWh	–
Mileage	$\approx$ 60	1 km	–
EFC	10	$\approx$ 0.0025	$f(SOE)$, see Equation (5.2.1)
SOH	10	$\approx$ 0.36 %	$f(E, E_{\text{nom}}, SOE)$, see Equation (5.2.2)

peration), charging (externally by AC or DC), or parking. The EFC and the SOH are calculated offline as part of the data cleansing:

$$EFC(t) = \left\lceil \frac{\sum_{t_0}^{t-1} |SOE(i+1) - SOE(i)|}{200\%} \right\rceil \tag{5.2.1}$$

$$SOH(t) = \frac{E_{\max}(t)}{E_{\text{nom}}} = \frac{E(t)}{SOE(t)} \cdot \frac{1}{E_{\text{nom}}} \tag{5.2.2}$$

with applying a symmetric weighted moving average and a manually-tuned time window of 10 d to smooth the SOH signal. This smoothed SOH value over EFC is shown as a boxplot of the whole fleet in Figure 5.2.

Inference from this plot to the average energy consumption of the fleet is not straight forward because the BEVs have different SOH values at their BOL and the start of data logging. Especially their SOH at BOL may be not 100 % because of the definition of E_{nom}. Also, for each BEV a different mileage span is logged during the logged time span. Thus, comparability to other SOH trajectories is not easily possible. At the beginning a sublinear decrease transitions to linear aging. After 400 EFC only few data is available, i.e., fleet SOH values need to be interpreted carefully there.

Sufficiency of the Sampling Time

When analyzing the sampling times of the signals given in Table 5.2, it is clearly visible in Figure 5.3 that a sampling time of 10 s is too high to accurately capture the discharge current signal of real-world vehicle operation. For example, an accelerati-

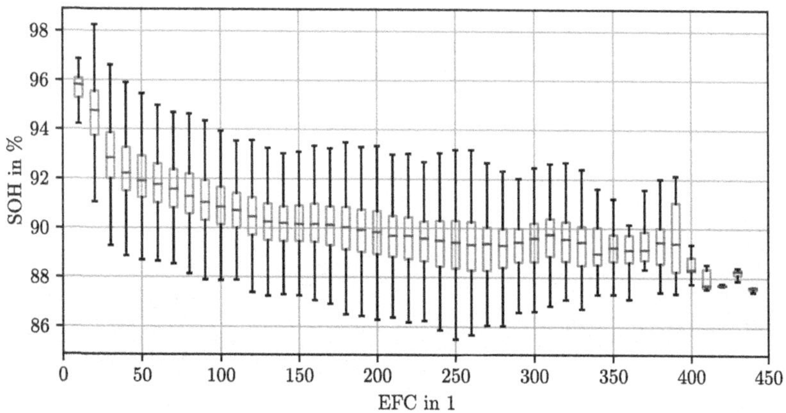

Figure 5.2 Fleet SOH curve

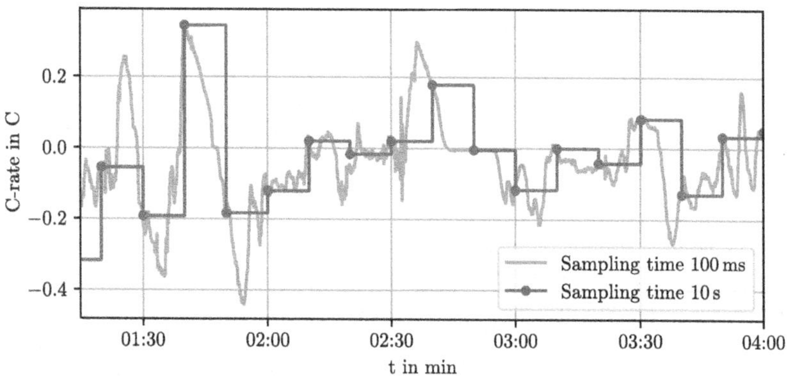

Figure 5.3 Exemplary comparison of the current signal with different sampling times

on phase of a few seconds after a stop at city traffic lights with different characteristic current profiles depending on the driver would not be captured accurately. For CC charging such a higher sampling time is less critical. Also, for SOE and temperature a smaller sampling time is sufficient because they are both less dynamic, i.e., only change slowly in a period of time of about up to 100 s.

Greenbank et al. [150] used histogram-based features for SOH forecasting as well. They concluded that increasing the sampling time of the two applied data sets [69; 70] from 10 s to 100 s resulted in no difference of performance [150].

However, the two data sets used identical CC discharging (I-CC) so only little information was lost by the increased sampling time. This limits the applicability of their finding for this data set (NI-VC).

Considering the histogram-based stressor data proposed in Chapter 6, the quality of approximation of the used histograms based on signals with a lower sampling time by a histogram based on signals with a higher sampling time may limit the suitability of the data set for SOH forecasting. This applies especially to the discharge current signal. As depicted in Figure 5.4, approximating discharge current histograms based on 100 ms sampling time with histograms based on a 10 s sampling time seems feasible, even if deviations exist. Notably during driving-mode at C-rates around zero the largest deviation exists. However, those very low C-rates have no substantial effect on battery aging. Still, all deviations mark a limitation of the used data set.

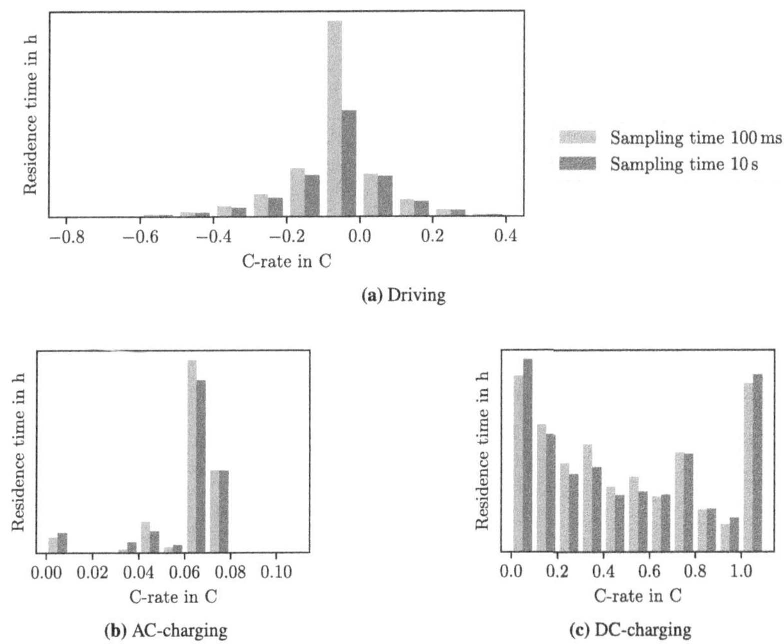

Figure 5.4 Approximation of histograms with high and low sampling time

Battery Cell State of Health Forecasting 6

The method and results that are presented throughout this chapter are primarily based on the previous publication [1].

6.1 Method

As mentioned in Section 3.1, battery aging is perceived as a state change from a current $SOH(t_1)$ to a future $SOH(t_2)$ due to aging causes. The aging causes are encoded in the battery operational data which consist of multidimensional time series signals of T, I and SOC. Thus, to fulfill key criterion 1a these T, I and SOC during the time span $\tau = [t_1, t_2]$ need to be present in the model input.

Thus, RNNs as a common ML model architecture to process time series data require consideration. However, Lin et al. [330] state that the effectiveness of RNNs and LSTMs on raw time-series cycling data still needs to be proven on field data. Also, RNNs suffer from vanishing gradients when trained on long time-series as mentioned in Section 2.2.1.2. Overall, RNNs struggle on coping with the different timely granularities in the battery time series data: On the one hand, the degradation of the SOH is usually long-term and slow. On the other hand, the operational time series data such as T, SOC, and I, the latter especially during discharge in BEVs, are usually more dynamic and of short-term nature. Thus, RNNs are not considered. Instead, as motivated for key criterion 1b in Section 3.5 data aggregation is desired.

For this task, the stressor extraction is the first part of the presented SOH forecasting method in Figure 6.1. As second part of the method, the extracted stressor

Supplementary Information The online version contains supplementary material available at https://doi.org/10.1007/978-3-658-43188-4_6.

F. von Bülow, *A Data-Driven Fleet Service: State of Health Forecasting of Lithium-Ion Batteries*, AutoUni – Schriftenreihe 170,
https://doi.org/10.1007/978-3-658-43188-4_6

data is input of an ML model, that outputs the state change ΔSOH from a current $SOH(t_1)$ to a future $SOH(t_2)$. The SOH values are assumed to be known for the training data. Nevertheless, in a production phase, only the current SOH at t_1 and the stressor data during τ needs to be known. The stressor data can either be derived from the operational data or be adapted based on a scenario as proposed in Section 3.3.

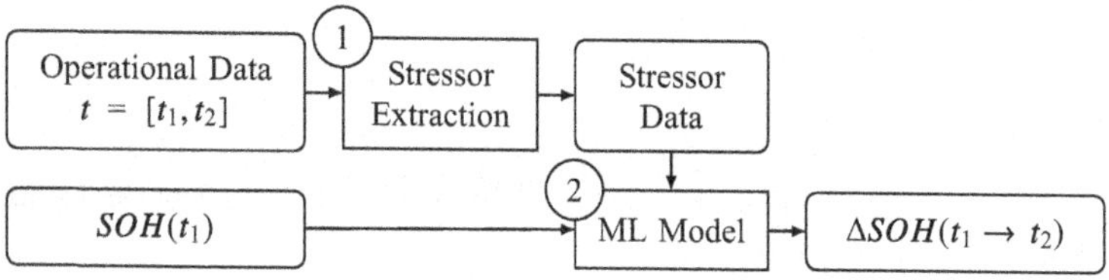

Figure 6.1 SOH forecasting method structure: (1) Stressor extraction (Section 6.1.1) and (2) ML model (Section 6.1.3).

Section 6.1.1 to 6.1.3 explain the stressor extraction, the signal selection, and the ML regression model respectively in more detail.

6.1.1 Stressor Extraction

In the stressor extraction, battery operational data is used to extract data of battery stressor types which are known to induce battery aging. A battery stressor type is defined by one or several relevant battery signals each limited by an interval (histogram-like binning). For example, one stressor type is the battery operation within a C-rate of 3C to 4C and a temperature of 31 °C to 32 °C as marked in the plot in Figure 6.2a. Figure 6.2b indicates that the battery has been operated 0.1036 h in the exemplary operational range during the depicted cycle. The corresponding stressor value is defined by Equation (6.1) as time elapsed when the battery is operated within the operational states of the stressor type. This corresponds to a two-dimensional extraction of [392]. When having several battery stressors, stressor tables like in Figure 6.2b can be formed, capturing all relevant battery stressor types.[1]

[1] These stressor tables are histograms which are called load spectrum or load collective in the field of mechanical fatigue analysis of machines and materials [441].

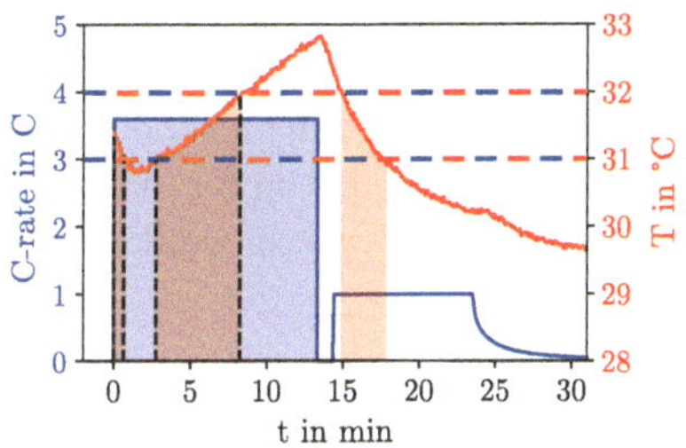

(a) Time series of current and temperature

		C-rate in C			
	t in h	$t_{0<I\leq 2}$	$t_{2<I\leq 3}$	$t_{3<I\leq 4}$	...
	...	...	...	...	...
T in °C	$t_{30<T\leq 31}$	0.3189	0	0.0347	...
	$t_{31<T\leq 32}$	0.0473	$4.9 \cdot 10^{-5}$	0.1036	...
	$t_{32<T\leq 33}$	0.0111	0	0.0838	...
	...	...	...	...	...

(b) Corresponding stressor table

Figure 6.2 Exemplary stressor extraction from time series data with C-rate and temperature T to form a stressor table which serves as model input. The signal interval width is medium according to Table 6.2. "I & T" according to Table 6.3. Data from battery cell no. 1 of the 50th cycle from [70].

$$t_{3<I\leq 4,\,31<T\leq 32} = \int_{t_1}^{t_2} f(\tau)\,\mathrm{d}\tau \text{ with } f(\tau) = \begin{cases} 1 & \begin{matrix}(3<I(\tau)\leq 4)\cap\\(31<T(\tau)\leq 32)\end{matrix} \\ 0 & \text{else} \end{cases} \tag{6.1}$$

Other work [309, p. 41] does not use absolute time values, but the relative time spent in a given bin. The authors argue that this provides more flexibility for variable time spans. In this work, min-max-normalization is applied as later described in Section 6.2 so that the model inputs are relative values as well. The stressor tables are finally vectorized, i.e., flattened, to have a suitable input shape for the MLP.

The stressor extraction as a type of data preprocessing has hyperparameters. These are the signal interval width, window width (w_{w}), and window shift (w_{s}). First, the signal interval width for the stressor types used in Equation (6.1) and displayed in Figure 6.2 is defining the sampling granularity of the signals' bins. For example, the temperature signal could be split into intervals of 1 °C or 2 °C. On the one hand, a finer signal interval width increases the number of input features as the stressor table in Figure 6.2 has more values, but will also provide more information to the ML regression model. On the other hand, a wider signal interval width increases the robustness of the method because noisy signals more likely stay in the correct interval. For the upper and lower bound of each signal, the maximum and minimum value of I, SOC, and T are used so that the complete range of these signals of the used data set is captured. For SOC this is easy as it is within 0 % to 100 % per definition. But for T and I it also depends on the battery operational domain.

Second, one training sample is defined by the two points in time t_1 and t_2 which correspond to the current $SOH(t_1)$ and future $SOH(t_2)$ respectively. However, the stressor table in Figure 6.2 is only based on a single charging cycle, so t_1 and t_2 are close. Thus, windowing is applied to the battery cycles defined by the cycle w_W which defines the number of cycles between t_1 and t_2 as shown in Figure 6.3. This means the stressor tables of the single cycles are simply added up for the whole window. The cycle w_W defines for how many future cycles the model learns to forecast the SOH.

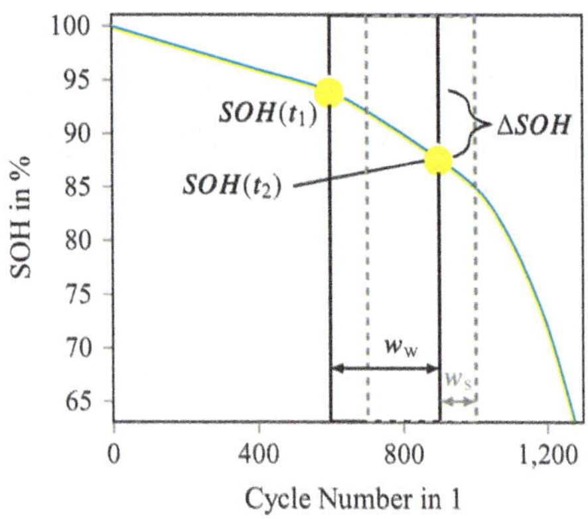

Figure 6.3 Hyperparameters of stressor extraction: Window width (w_W) and window shift (w_S).

Third, to extract further training samples the cycle window is shifted by w_S as visualized by the dashed gray box in Figure 6.3. This procedure is known as sliding window approach in the domain of time series data preprocessing with the w_W and w_S called horizon and stride respectively [442]. The samples extracted from the time series of a single battery are defined by their start and end cycle number $k_{\text{start},n}$ and $k_{\text{end},n}$ respectively:

$$\{(k_{\text{start},n}, k_{\text{end},n})\} = \{((n-1)\cdot w_S + 1, (n-1)\cdot w_S + w_W)\} \text{ with } n = 1, 2, 3, \ldots$$
$$\text{e.g., with } w_W = 50,\ w_S = 25\text{: } \{(1, 50), (26, 75), (51, 100), \ldots\} \quad (6.2)$$

with $k_{\text{end},n} \leq k_{\text{cycle life}}$ up to the end of the available time series data of the battery's cycle life. The EOL of the battery is in most cases reached after a number of cycles that is not $w_S \cdot n + w_W$ with $n \in \mathbb{N}$. To avoid shorter windows than w_W at the EOL, the last window is until the last cycle (EOL) and begins by w_W before the

EOL. Exemplary training samples consisting of the current $SOH(t_1)$ and vectorized stressor tables are shown in Table 6.1.

Table 6.1 Exemplary part of the input feature vector x_n and output value y_n. The signal interval width is coarse according to Table 6.2. $w_W = 50$, $w_S = 50$. Data from [69].

			Input x_n					Output y_n
	Sample n	t in h	Feature 1	Feature 2	Feature 3	Feature 4	...	Label
		Cycles	$SOH(t_1)$	$t_{0<I\leq 3,\ 0<T\leq 28}$	$t_{0<I\leq 3,\ 28<T\leq 31}$	$t_{0<I\leq 3,\ 31<T\leq 34}$	...	ΔSOH
Cell 1	1	1–50	97.33 %	0 h	15.27 h	2.73 h	...	0.55 %
	2	26–75	97.87 %	0 h	14.04 h	2.70 h	...	−0.02 %
	3	51–100	97.88 %	0 h	14.1 h	2.6 h	...	−0.07 %
	...	...	...	...	...	...	...	...
	47	1216–1175	93.80 %	0 h	10.93 h	3.51 h	...	−0.35 %
Cell 2	48	1–50	97.75 %	0 h	14.16 h	2.56 h	...	0.58 %
	49	26–75	98.37 %	0 h	14.49 h	42.24 h	...	−0.07 %
...	...	...	...	...	...	...	...	...

w_S controls the number of generated samples and the amount of new operational data in each sample as the window shifts forward through time. If w_S is larger than w_W, some sections in the degradation curve would be skipped, leading to a loss of information. For example, in the case of $w_S = 50$ and $w_W = 25$, the generated windows are exemplarily from cycle 1–26 and 51–76, but the cycles 27–50 would not be used. Thus, in this case, $w_S = w_W$ is chosen to avoid skipping any data or avoid it completely by the setup of this design of experiments (DOE). If w_S equals w_W, each sample contains entirely new operational data. If w_S is smaller than w_W, the windows are partially overlapping. When increasing w_W and w_S, the quantity of training samples decreases as fewer cycle windows can be generated from one battery.

No down sampling is applied to the time series data. A sampling time is sufficiently high if it captures short spikes, especially of the current. This is assumed to be the case in the ECUs and the BMS anyways. Hence, no additional down sampling is considered before aggregation of the time series data to stressor tables. No additional noise is added to the signals because, first, it is assumed that the precision of the signals used by the BMS is also sufficient for the presented method. Second, the method is robust to noise in the time series of current, temperature, and SOC because of the binning in the stressor tables.

Applying the wavelet transformation to time series data to transform it from the time to the frequency domain [443] was beneficial in other industrial ML applications [444]. However, for this application different window sizes lead to different output shapes of the wavelet transformation. Furthermore, for operational data from laboratory with CC steps of c and identical current profile over cycles (I-CC) from Section 2.1.6.2 only very few or only one frequency would be obtained, i.e., CC can be perceived as oscillation with frequency spectrum $c \cdot \delta(f)$.

6.1.2 Decision on Signal Selection

This section discusses the selection of the ΔSOH as output as well as the C-rate and SOC as input signals.

Output: SOH vs. ΔSOH

Figure 6.1 indicates that the model output is the future ΔSOH. As proposed in Section 3.1, an alternative model output would be the $SOH(t_2) = SOH(t_1) + \Delta SOH(t_1, t_2)$ or correspondingly the capacity fade ΔC. In the related work in Table 4.3, both the capacity C and the relative SOH also as Δ-values are applied.

The relative SOH is chosen instead of the absolute capacity C to facilitate comparability and transferability for batteries with different C_{nom} as done by [445].

Second, the ΔSOH is chosen for several reasons: Learning the identity for little or no aging is hard for an ML model and skip connections would likely be needed [446]. It is assumed to be much easier to learn the ΔSOH and do the addition of ΔSOH and $SOH(t_1)$ explicitly afterwards, if required, compared to force the ML model to pass the $SOH(t_1)$ somehow through the ML model until the output. Furthermore, it enables the ML model to learn a similar distribution of output values for different aging trajectories.

Also, in the domain of time series forecasting learning a Δ-value is common because it enables more consistent and smooth forecasts so that not only a single forecast step, but the complete forecast is sensible. The same applies for the prediction in technical systems, e.g., the prediction of the temperature change is preferred over the temperature [447].

Input: C-rate vs. current

In the context of this method, the symbol I refers to the C-rate, i.e., the current relative to C_{nom}, not to the absolute current. The current expresses the flow of electrical charge per second. Normalization by C_{nom} relates the electrical charge not only to time, but also to the nominal available amount of cyclable lithium inventory which is affected by the degradation mechanisms Thus, as for the SOH compared to the absolute capacity, this enables better comparability and transferability among batteries with different C_{nom} like [448, p. 6] do as well. Furthermore, when integrating the C-rate over time the number of EFC is directly obtained.

Input: SOC vs. voltage

The SOC is favored over the voltage because it is more meaningful and interpretable regarding battery operation for non-battery experts like customers and BEV fleet managers than the voltage. This is especially the case when applying scenario data to a trained model as suggested in Section 3.3. Furthermore, the voltage is more dynamic given its proportionality to the dynamic current by Ohm's Law, while the SOC can be perceived as integral of the current. Then again the violation of the voltage limits may accelerate irreversible aging as shown in Figure 2.5.

6.1.3 Machine Learning Regression Model

The ML model uses the current $SOH(t_1)$ and the stressor data as features according to Table 6.1. Once a model of a certain battery type has been trained, only a minimal amount of training data shall be required to obtain a model for a new battery type (Key criterion 2). Thus, an ML model suitable for transfer learning is necessary. Therefore, MLPs are chosen as they store different information on different layers as introduced in Section 2.2.1.1, which allows adaption to similar problems via transfer learning. Only for comparison to [177; 392], also GPR is implemented in Experiment 4. Furthermore, MLPs are a well-established ML architecture. Another publication [4] in the context of this thesis which is not discussed further here, also examined decision tree (DT), random forest (RF), eXtreme Gradient Boosting (XGBoost), histogram-based gradient boosting (HistXGBoost), multilayer perceptron with residual blocks (ResMLP), and convolutional neural network with residual blocks (ResCNN) on the task of SOH forecasting.

6.2 Design of Experiments

For developing this method for SOH forecasting it is essential to find suitable hyperparameters for the stressor extraction described in Section 6.1.1. The examined hyperparameters are the signal interval width for current, temperature, and SOC as well as the cycle w_{w} and w_{s}. The objective is to find the right balance of input features and training samples. Further, different combinations of the signals of current, temperature, and SOC to stressor tables for charging ($I > 0$), discharging

($I < 0$), and hold mode ($I = 0$) are examined. In addition, the capability of the SOH forecasting model to generalize is investigated.
In this chapter, the battery cell aging data set DD [70] from laboratory operation presented in Section 5.1.1.1 is used.

Experiment 1a: Single window width w_W
The cycle w_W determines how many cycles into the future the model can forecast the SOH following Figure 6.3. To examine how well the model forecasts small and big SOH degradations, different w_W are defined. The maximum w_W is limited by the battery cell no. 2 of the DD data set which ages the fastest. Thus, data sets are created with w_W of 25, 50, 100, 200, 400, and 530 cycles (W1 to W6).

Experiment 1b: Combined window width w_W
Expectedly, a model trained with a data set of a certain w_W will only be suitable to predict on samples of that or a similar w_W. This would limit the practical applicability of the model for example for fleet managers because they may be interested into the aging after different amounts of cycles using one model. Thus, the aforementioned w_W are combined to further data sets as follows: $w_W \in \{\{25, 50\}, \{50, 100\}, \{25, 50, 100\}, \{200, 400, 530\}, \{50, 200, 530\}, \{50, 200, 400, 530\}\}$ (W7 to W12). However, different signal interval widths cannot be combined easily as they have different input feature shapes requiring different input layers.

Experiment 1c: Window shift $w_S = 50$
A small w_S will improve the ability of the MLP to interpolate as more samples are available and the size of the training data will increase as well. When applying the method, e.g., in BEVs the stressor tables may not be transmitted and saved in a central cloud after every battery cycle, but only after certain intervals of time or EFC. Thus, data sets for experiments 1a and b have $w_S = 25$ and for experiment 1c $w_S = 50$ (see Figure 6.3). Further, the data sets W1 to W12 each contain either fine, medium, or coarse signal interval widths as displayed in Table 6.2 (F, M, and C respectively). Thus, 36 data sets for each w_S are created. These data sets consist of stressor tables using two combined signals (2D) in the operational modes of charging, discharging, and hold as shown in Table 6.3 (2D stressor table, variant A).

Table 6.2 Signal interval width for current, temperature and SOC

Signal interval width	Current	Temperature	SOC
Fine (F)	0.5C	0.5 °C	5 %
Medium (M)	1C	1 °C	10 % at 0 % and 100 %, else 20 %
Coarse (C)	3C	3 °C	20 %

First experiments deepened in Section 6.3 show that the coarse signal interval width (C) with w_S of 25 and w_W of 200, 400, and {25, 50, 100} cycles (W4, W5, and W9 each with C-2D, variant A) result in a coefficient of determination (R^2) greater than 0.99. Thus, further data sets are created to examine variations of the stressor tables for w_W of W4, W5, and W9 as shown in Table 6.4. These all have a coarse signal interval width.

Table 6.3 Combined signals for 2D stressor tables, variant A and B

	Charging	Discharging	Rest
Variant A	T & SOC I & SOC I & T	T & SOC I & SOC I & T	T & SOC
Variant B	T & SOC I & SOC	T & SOC I & SOC	T & SOC

Table 6.4 Data sets created for the experiments with DD data. For the specification of the signal interval width see Table 6.2 and for the stressor tables Table 6.3.

Exper. no.	Abbreviation	Signal interval width	w_W	w_S	Stressor tables
1	2D, Variant A	F, M, C	W1-12	25 and 50	A
2	2D, Variant B	C	W4, W5, W9	25	B
3	3D	C	W4, W5, W9	25	A
4	2D, Variant A	C	1 cycle (W13)	1	A

Experiment 2: 2D stressor tables, variant B

Second, a data set with 2D stressor tables is created, but with adapted signal combinations as shown in Table 6.3 (Wx-C-2D, variant B). In comparison to variant A in Experiment 1, this variant B does not contain any explicit stressor combination of current and temperature. This shall ease the applicability of the model as the dependency of current and temperature does not need to be considered explicitly when designing a sample for a forecast by a human (Part of key criterion 1b).

Experiment 3: 3D stressor tables
Third, as stated in Section 2.1.4, current, temperature, and SOC are relevant signals inducing battery aging. To examine them as joint stressors, a data set is created with 3D stressor tables combining current, temperature, and SOC over all battery operational modes (Wx-C-3D).

Experiment 4: Window width $w_{\mathrm{w}} = 1$ and window shift $w_{\mathrm{s}} = 1$
Forth, Richardson et al. [392] use a data set with $w_{\mathrm{w}} = w_{\mathrm{s}} = 1$. For comparison with them, further a data set with $w_{\mathrm{w}} = w_{\mathrm{s}} = 1$ is created using the 2D stressor table, variant A (W13-C-2D, variant A).

For each data set a corresponding hyperparameter optimization was executed using Hyperopt version 0.2.4 with TPE which resulted in a final model for each data set [267; 268]. Hyperopt parameters were set to maximum of 1000 evaluations and 100 random startup evaluations. Each data set was split for training, validation, and test by the ratio of 80:10:10. Min-max-normalization is applied to each data set. The MSE on the validation data set was set as optimization metric for the hyperparameter optimization. The variable hyperparameters of the ANN and their range are chosen form experience and shown in Table 6.5. The constant hyperparameters of the ANN are the optimizer adaptive moment estimation (Adam), 80 epochs, MSE as loss function, maximum of ten hidden layers and a linear activation function of the output layer. Version 2.0.4 of TensorFlow was used as back-end including version 2.2.4 of Keras.

Table 6.5 Overview of the used hyperparameters of the MLP.

Hyperparameter		Values
Activation function of hidden layer [250]		Sigmoid, ReLU, tanh
Batch size		32, 64, 128
Learning rate α		0.01, 0.001, 0.0001
Number of layers		[1, 10], in steps of 1
Neurons per layer		[5, 200], in steps of 10
Choose one, others are 0:	Regularization parameter λ_1 or λ_2 [247; 249; 256]	0.01, 0.001, 0.0001
	Dropout rate [449]	0.1, 0.2, 0.5

6.3 Results

First, the results of the 36 models trained on data sets with $w_W = 25$ are evaluated (Experiment 1a and b). Afterwards, the results of $w_S = 25$ and $w_S = 50$ are compared with each other (Experiment 1c). Then, Experiments 2 & 3 analyze the adapted 2D and 3D stressor tables respectively. Further, Experiment 4 assesses $w_W = w_S = 1$. Finally, MLPs trained with 2D stressor tables each with $w_S = 1$ and $w_W = 25$ as well as a GPR fitted on the last data set are evaluated.

The hyperparameters of four MLPs are displayed in Table E2. The hyperparameter optimization never converges choosing the maximum of the value range for the number of layers and neurons per layer in Table 6.5. This underlines a suitable choice of the value range of the hyperparameters.

As shown in Appendix E.1 (accessible in the Electronic Supplementary Material), data sets with a large w_W have a higher mean ΔSOH, i.e., a higher mean output value. The variance of the output values also increases resulting in a greater variance of ΔSOH the models needs to fit. Thus, when comparing models with different w_W, i.e., a different output value distribution, the coefficient of determination (R^2) is suitable because it normalizes the MSE by the variance σ_y^2 as shown in Equation (2.2.8) [274]. However, when comparing models with the same w_W the RMSE is still favorable as it is non-relative, thus, indicating the error in the real output value unit.

Experiment 1a: Single window width w_W

For each data set W1 to W12, Figure 6.4 shows a positive influence of a finer signal interval width on the models' RMSE. A finer signal interval increases the number of features as shown in Appendix E.1. This improves the distinguishability of the samples and thus reduces the RMSE. This effect seems to diminish from W1 to W6 in Figure 6.5 when looking at R^2. However, as noted earlier, R^2 is an unsuitable metric for the analysis of the signal interval width due to its normalization. The described effect can still be observed when zooming into Figure 6.5.

Despite the use of regularization techniques, overfitting occurs for all models without combined w_W in Figure 6.4 (W1 to W6). For these models the RMSE on the test data set is on average 65 % worse than on the training data set. It has to be noted that the RMSE from W1 to W6 also increases because the mean of the output value increases (see Appendix E.1). Thus, in the following R^2 is used as a relative RMSE (see Equation (2.2.8)).

When increasing w_W for the coarse signal interval width, measured by R^2 in Figure 6.5 the models fit better. For W4 to W6 in Figure 6.5 for all three data sets R^2 is above 0.99, meaning these models explain at least 99 % of the variance in the output data. For these models, R^2 on the training data set is on average 0.49 % better than on the test data set. For W1 to W3, this value is 2.05 %. Thus, the generalizability is better for W4 to W6 than for W1 to W3. One reason for this might be the data set size which decreases from W1 to W6 (see Appendix E.1). The more important reason is seen in the output value distribution also shown in Appendix E.1. From W1 to W6 the mean of the output values increases which means the model is trained to learn greater SOH degradations. The variance of the output values also increases which means that the model learns a greater variety of SOH degradations. Both ease the distinguishability of the samples for the model.

From this observation, it is concluded that a data set with samples that covers a longer w_{w} is more suitable for model training. A longer w_{w} also makes sense for the practical applicability for fleet operators: They are more interested in the battery aging after a longer period like more than 200 cycles (W5). They are not so much interested in the battery aging after only 25 cycles (W1), similar to the proposal of [60, p. 6]. Assuming an intensive use of the BEVs with two charging events per day, these two w_{w} would correspond to a forecast of three or only 0.5 months.

Experiment 1b: Combined window width w_{w}

The combination of different w_{w} to one data set (W7 to W12) increases the data set size as shown in Appendix E.1. When combining different w_{w} for the coarse signal interval width, different effects are observed: Combining W1 and W2 to W7 does not improve R^2 on the training data, but improves the generalization on validation and test data. Remarkably, separately for W1, W2, and W3 R^2 is 0.9714 on average for the test data, but when combining W1, W2, and W3 to W9 R^2 is 0.993 on the test data. Also, W9 achieves a higher R^2 compared to W7 and W8 on the test data. Thus, it can be concluded that combining data sets with short w_{w} which are similar to each other does not provide enough incentive for generalization. Improved generalization with combined w_{w} is also observed for W10 to W12, though the effect decreases for combining longer w_{w}. The models W7 to W12 generalize better because the models have to abstract features from samples with different w_{w}. Thus, combining different w_{w} into one data set is favorable.

Experiment 1c: Window shift $w_{\mathrm{s}} = 50$

After examining models trained on data sets with $w_{\mathrm{s}} = 25$, the same configurations only with $w_{\mathrm{s}} = 50$ are examined. The results for R^2 and RMSE show slightly inferior results for all models (see Table E.4, accessible in the Electronic Supplementary Material). When increasing w_{s} from 25 to 50 the number of samples is roughly halved. With fewer samples, the models were neither able to fit nor to generalize as well as they did with $w_{\mathrm{s}} = 25$. Furthermore, the single samples have more overlapping cycles with $w_{\mathrm{s}} = 25$ which leads to better training results without the necessity for more battery time series data. Also, the MLP has more sample points to interpolate in between.

Experiment 2: 2D stressor tables, variant B

The data sets with coarse 2D stressor tables variant B have 30 features less than with 2D stressor tables variant A. Switching from 2D stressor tables variant A to B did not worsen the RMSE or R^2 of the models, as shown in Table 6.6. Thus, the combination of battery signals into 2D stressor tables with variant B seems sufficient for learning SOH degradation.

Table 6.6 Results of models trained on 2D stressor tables, variant B, and $w_s = 25$ (Wx-C-2D, variant B).

Interval	w_w	RMSE			R^2		
		Train	Val	Test	Train	Val	Test
Coarse	200 cycles (W4-C-2D, variant B)	0.1538	0.208	0.3034	0.9987	0.9966	0.9897
	400 cycles (W5-C-2D, variant B)	0.1978	0.1794	0.3186	0.9989	0.9976	0.9946
	{25, 50, 100} cycles (W9-C-2D, variant B)	0.1321	0.1723	0.1568	0.9947	0.9911	0.9935

Experiment 3: 3D stressor tables

The data sets with coarse 3D stressor tables combining current, temperature, and SOC over all battery operational modes (W4-/W5-/W9-C-3D) have 20 more features than with 2D stressor tables, variant A. The results of models trained on coarse 3D stressor tables and $w_s = 25$ are shown in Appendix E.1. Switching from 2D to 3D stressor tables did not improve the RMSE or R^2 of the models significantly, as shown in Tables 6.6 and 6.7. Thus, the combination of battery signals into 2D stressor tables seems sufficient for learning SOH degradation.

Experiment 4: Window width $w_w = 1$ and window shift $w_s = 1$

The features used by Richardson et al. [392] correspond to $w_w = w_s = 1$ in the presented method. Their best validation RMSE for predicting the capacity change of a battery cell is 0.0201. For the comparison with them, the coarse data set with $w_w = w_s = 1$ and 2D stressor tables variant A from this chapter is used. On this data set, a test RMSE of 0.0357 is achieved. Compared to [392], the method from this work performs worse by 43 %, but [392] did not provide the RMSE on the training data set. Thus, potential overfitting of their model cannot be checked. The model from this work has an RMSE on the test data set which is only 9 % worse

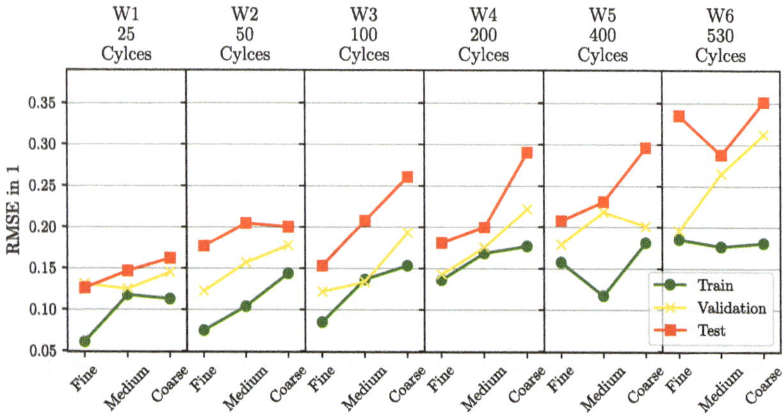

(a) Model performance with single w_w

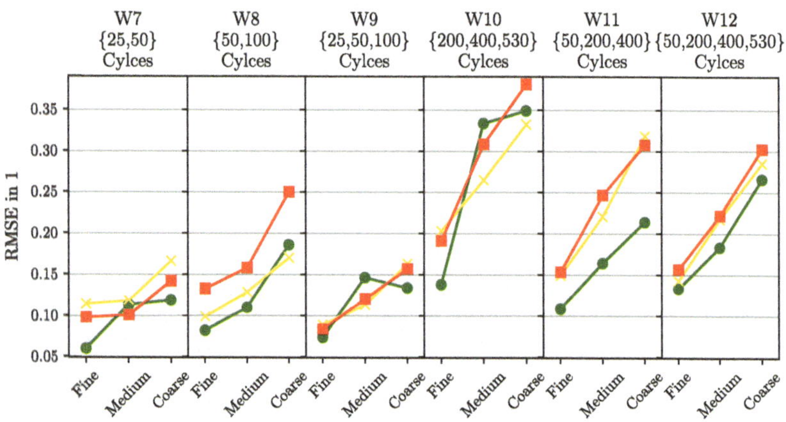

(b) Model performance with combined w_w

Figure 6.4 Model performance W1 to 12 with $w_s = 25$ measured by RMSE.

than on the training data set. Further, [392] did not publish R^2 of their models, but on the aforementioned data set the test RMSE of 0.0357 corresponds to R^2 of only 0.388, as shown in Table 6.8. The worst test R^2 of all other examined variations shown in Figure 6.5 is 0.949. Concluding R^2 of $w_w = w_s = 1$ is 59 % worse than the worst of the other w_w (W1 to W12).

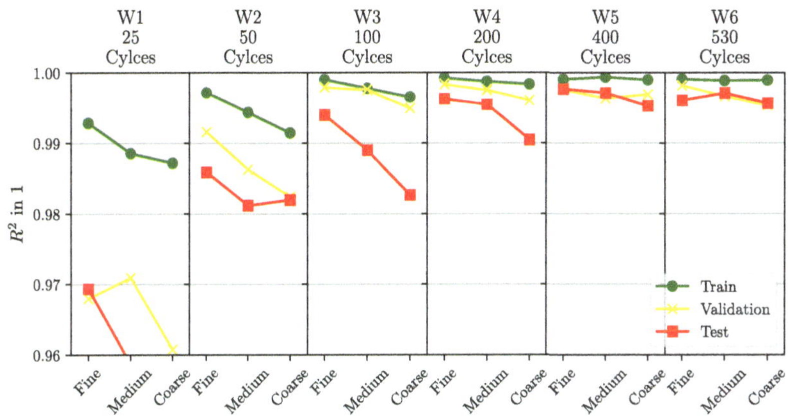

(a) Model performance with single w_w

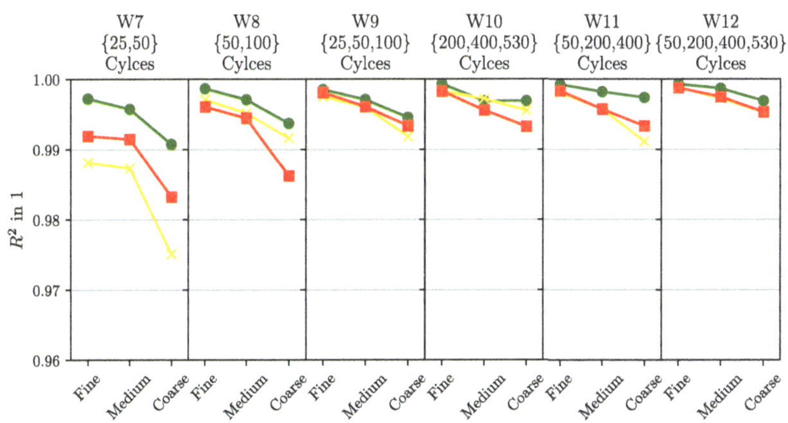

(b) Model performance with combined w_w

Figure 6.5 Model performance W1 to 12 with $w_s = 25$ measured by R^2.

This becomes also clearly visible in Figure 6.6 which depicts the model fit of three different data sets each split into training, validation, and test data by comparing the predicted versus measured ΔSOH values. A good fit would show predictions that are close to the true values. Thus, the predictions (dots) should be closely around the dashed identity line. The mentioned data set with $w_w = w_s = 1$ is shown in Figure 6.6a. Not even on the training data set a good fit is reached. In addition,

Table 6.7 Results of models trained on 3D stressor tables and $w_s = 25$ (Wx-C-3D).

Interval	w_w	RMSE			R^2		
		Train	Val	Test	Train	Val	Test
Coarse	200 cycles (W4-C-3D)	0.2159	0.234	0.2845	0.9981	0.9955	0.9901
	400 cycles (W5-C-3D)	0.2193	0.2507	0.3828	0.9981	0.9952	0.9922
	{25, 50, 100} cycles (W9-C-3D)	0.1226	0.1728	0.1581	0.9954	0.9911	0.9933

Table 6.8 Results of models trained on 2D stressor tables, variant A, and $w_w = w_s = 1$ (W13-C-2D, variant A).

Interval	w_w	RMSE			R^2		
		Train	Val	Test	Train	Val	Test
Coarse	(W13-C-2D, variant A)	0.0325	0.0312	0.0357	0.5499	0.4436	0.3881

the overfitting is evident as the deviation from the identity line is even larger for validation and test data. Contrarily, the data set W9 with $w_s = 25$ and 2D stressor table, variant A, depicted in Figure 6.6b shows a very good fit and generalization which are important for practical application. Richardson et al. [392] use GPR with a Matérn-Kernel with $\nu = 2.5$. When using the same data as in Figure 6.6b, but using GPR as [392] instead of MLP as in Figure 6.6b, the results shown in Figure 6.6c were obtained. Compared to Figure 6.6b, the overfitting is even higher. Compared to Figure 6.6b, the RMSE on the training data is better, but overfitting occurs. The comparison with Figure 6.6b indicates that GPR overfits more easily on the same data. Thus, MLPs seem preferable.

[392] apply a specialization of the presented stressor extraction with $w_w = w_s =$ 1 cycle. When applying their model for multiple cycles in a loop, propagation of the model error is expected as discussed in Section 3.5.

6.4 Summary and Contribution to the Research Questions

Given the limitations of the related work on SOH forecasting outlined in Section 4.4, this chapter proposed an ML method for SOH forecasting applicable for BEV fleet managers and battery designers in real-world applications (RQ.1). BEV fleet managers can improve the operation and replacement of their fleet members by applying the proposed SOH forecasting model. The model inputs are the battery operation time within certain operational ranges defined by combinations of the battery signals I, SOC, and T during a certain number of cycles.

Summarizing the results, combining different w_w to one training data set improves generalization. Also, models trained with features extracted for fine signal

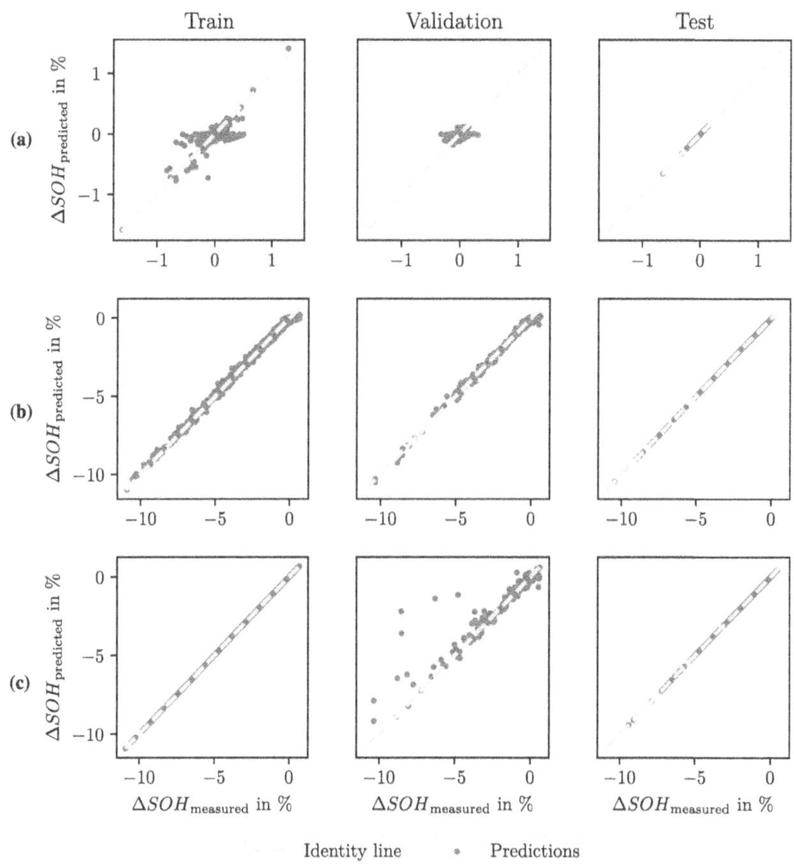

Figure 6.6 Predicted versus measured ΔSOH values for:
6.6a) MLP, $w_w = w_s = 1$ (W13-C-2D, variant A);
6.6b) MLP, $w_w = \{25, 50, 100\}$ (W9-C-2D, variant A), $w_s = 25$;
6.6c) GPR, $w_w = \{25, 50, 100\}$ (W9-C-2D, variant A), $w_s = 25$ (Same data as in 6.6a) but with GPR);
Common data set characteristics: Coarse, 2D stressor table, variant A.

intervals provide more precise forecasts than coarse signal intervals. $w_s = 25$ is favorable to $w_s = 50$. Using 3D stressor tables does not improve the model's forecast compared to 2D stressor tables, while 2D stressor tables variant B compared to variant A of the 2D stressor tables does not worsen the model's forecast. Furthermore, the fineness of the operational ranges of the signals does not limit the model if the w_w is larger than 100 cycles or different w_w are combined. Also, GPR was overfitting more easily than MLP on the same stressor data.

The method also enables the application of transfer learning with the possibility to reduce the size of the training data significantly, once an initial model for a battery type has been trained (Key criterion 2). This is further examined in the next Chapter 7.

7 Transfer of Battery Cell State of Health Forecasting

After showing the suitability of the proposed SOH forecasting method on battery cell data from laboratory operation without limited data availability in the previous Chapter 6, this chapter evaluates a transfer learning approach of the same method to enable quick applicability on new batteries (Key criterion 2) as motivated in Section 3.4.

In the following, this chapter examines "when" and "how to transfer". "When to transfer" concerns the data availability in the target domain, i.e., the number of samples and their distribution by the quantity and age of batteries. This shall enable practitioners to estimate the amount of data required for a successful transfer. "How to transfer" concerns the transfer method, i.e., freezing of layers of the pretrained source model.

The method and results that are presented throughout this chapter are primarily based on the previous publication [8].

7.1 Method

First, an SOH forecasting MLP model in D_S, the source model, is (pre-)trained according to Section 6.1. Second, the source model is parametrically transferred to another domain D_T. We assume D_S to be a battery type with plenty of training samples available. Further, D_T represents a new battery type that has little available training samples at the point in time t of the transfer ($|D_T(t)| \ll |D_S|$). We assume that in time the amount of training samples of the new batteries increases ($|D_T(t_1)| < |D_T(t_2)|$ for $t_1 < t_2$). The battery operational time series data is preprocessed

F. von Bülow, *A Data-Driven Fleet Service: State of Health Forecasting of Lithium-Ion Batteries*, AutoUni – Schriftenreihe 170,
https://doi.org/10.1007/978-3-658-43188-4_7

according to the previous publication presented in Chapter 6 [1]. For all data sets the same signal interval width and the same boundaries of the stressor tables in D_S and D_T are used so that $\mathcal{X}_S = \mathcal{X}_T$. However, for different operation and different battery types $P(X_S) \neq P(X_T)$ holds true because the distribution within the stressor tables and also $SOH(t_1)$ are different. As $P(X_S) \neq P(X_T)$, also the domains are different ($D_S \neq D_T$). Regarding the learning tasks, we assume $\mathcal{Y}_S = \mathcal{Y}_T$ because ΔSOH is always the single output. However, it holds that $f_S(\cdot) \neq f_T(\cdot)$ because the range of ΔSOH is different for the battery types in $\mathcal{T}_S$ and $\mathcal{T}_T$ given $X_S = X_T$. E.g., the target domain may be unbalanced with more new batteries. As $f_S(\cdot) \neq f_T(\cdot)$, also the source and target task are different ($\mathcal{T}_S \neq \mathcal{T}_T$) leading to the setting of inductive transfer learning. Thus, labeled data from the target domain are required to induce $f_T(\cdot)$ [275].

In the context of this thesis, "what to transfer" concerns the common characteristics of the relationship of current $SOH(t_1)$, battery operation (C-rate), temperature, and SOC) and the future $SOH(t_2)$, i.e., the function defined in Equation (3.1.2).

The following two sections examine "when" and "how to transfer". "When to transfer" concerns the data availability in the target domain, i.e., the number of samples and their distribution by quantity and age of batteries. This shall enable practitioners to estimate the amount of data required for a successful transfer. "How to transfer" concerns the transfer method, i.e., freezing of layers of the pretrained source model.

7.1.1 How to Transfer: Layerwise Freezing

Via a parametric transfer of the weights and biases the knowledge is transferred from the source model to the target model. As common procedure in computer vision shown in Section 2.2.5 and 4.3, weights and biases of the model's layers can be frozen, i.e., they do not change while further training. Thus, the knowledge learned from D_S is preserved in the frozen layers. In the unfrozen layers, the weights pretrained in D_S serve as starting point compared to randomly initialized weights. After freezing, training the target model is continued with the target training data. The parametric transfer has the transfer hyperparameter learning rate α, epochs, and number of frozen layers n_{frozen}. The number of frozen layers is counted from the input layer towards the output layer as the example in Figure 7.1 shows. For answering "how to transfer", the optimal selection of n_{frozen} is examined for achieving the best result on the target test data.

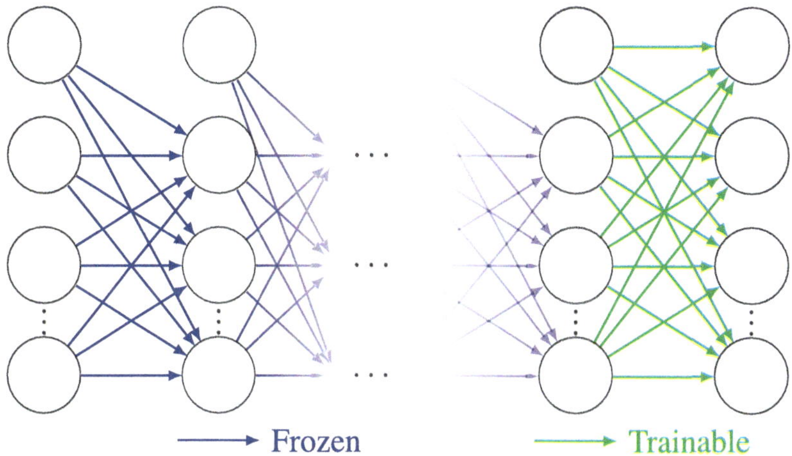

Figure 7.1 MLP with $L - 1$ potentially trainable layers and exemplary freezing of all layers except of the still trainable output layer, i.e., $L - 2$ frozen layers

7.1.2 When to Transfer: Data Availability

"When to transfer" for battery aging models refers to the temporal availability of target data. Thus, the target data is artificially split into training and validation data which is available at transfer time and test data which has not been recorded at transfer time. The test data will be input to the model when it runs in production, i.e., in prediction mode applied, e.g., by an automotive manufacturer or BEV fleet managers. Thus, the model evaluation on the test data is a suitable measure for the success of the transfer. The amount of training target data is measured by the number of samples available for the transfer. The distribution is defined by the quantity and age of batteries, i.e., the maximum cycle number.

The target data is split according to three scenarios inspired by an automotive manufacturer that will introduce a BEV with a new battery type. The new battery type is the target domain D_{T} from an ML point of view. For each data split, first the order of the battery cells is shuffled. Then samples are added to the training and validation data set according to the data split until the specified number of samples is reached. If multiple w_{W} are present in the data set, samples are sorted ascendingly. The remaining samples compose the test data set. The three data splits are shown exemplarily in Figure 7.2:

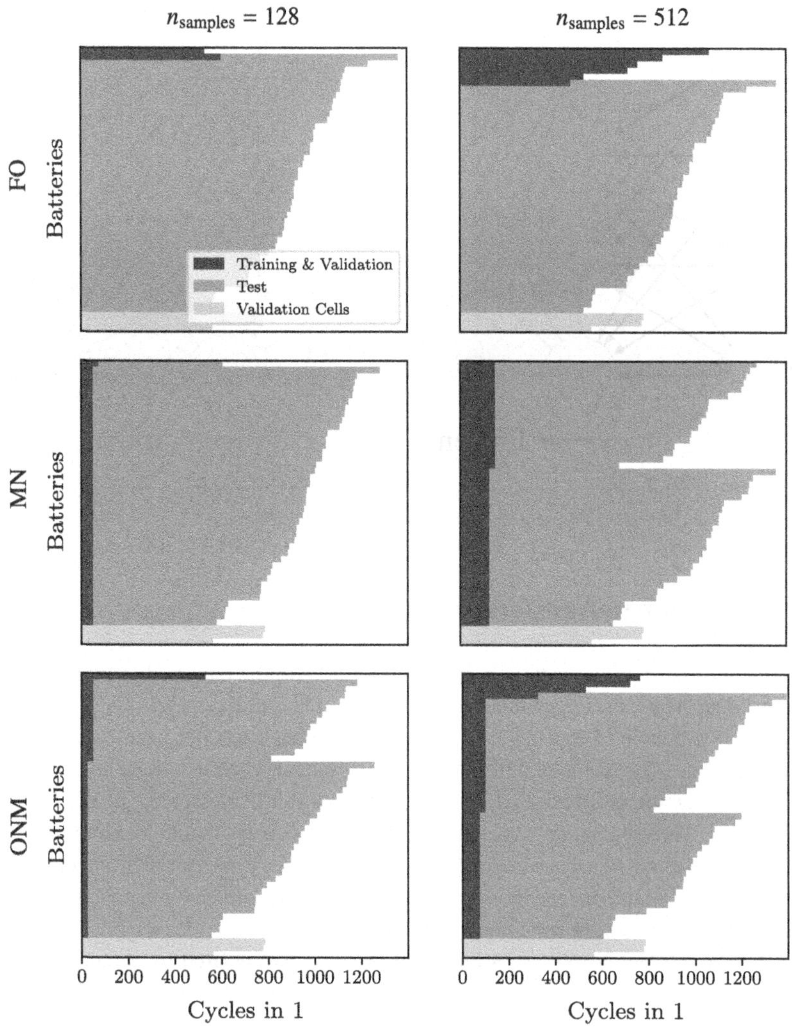

Figure 7.2 Exemplary data splits few old (FO), many new (MN), and old new mixed (ONM) with $n_{\text{samples}} \in \{128, 512\}$ using the CL data set. Each bar visualizes the allocation of the time series data of one battery to the data sets

1. The data split few old (FO) assumes that before the market launch of a new BEV type endurance tests are run: A few vehicles are intensively driven so that components like the batteries age quickly [450; 450]. Further, limited data from laboratory experiments is possibly available. Thus, target data of few, but old batteries is available for transfer learning.
2. The data split many new (MN) assumes that after the market launch of a new BEV type after some time already many vehicles are driven by customers. However, these batteries have only aged very little or non-measurably. Thus, target data of many, but new batteries is available for transfer learning.[1]
3. The data split old new mixed (ONM) assumes the availability of data from the first two data splits. Thus, target data of many, but new batteries and few, but old batteries is available for transfer learning.

7.1.3 Benchmarks

For comparing the performance of transfer learning with non-transfer learning approaches whilst having the same availability, Table 7.1 defines three benchmarks with a different composition of training and validation data. All benchmarks are tested on the same test data like the MLP with transfer learning according to the data split type and the number of samples available. Only the training and validation data differ. Source only (SO) is the source model without further modifications tested on the target test data. Source target mixed (STM) uses data from source and target domain for training and validation. The target only (TO) model is exclusively trained and validated on target domain data. For comparability, all benchmarks are MLPs with the same network architecture as the source model, but the weights are randomly initialized. Only their learning rate and batch size are optimized.

Table 7.1 Overview of benchmarks for transfer learning

Benchmark	Training and validation data	Test data
Source only (SO)	D_S	
Source target mixed (STM)	D_S & D_T	D_T
Target only (TO)	D_T	

[1] Cerqueira et al. [452, p. 2000] call this variant of a prequential approach growing window.

7.1.4 Data

An overview of the extensive data corpus aggregating five different known public data sets [68–70; 169; 211] regarding battery operation and materials is given in Table 5.1. These public data sets: 1. provide data of sufficient battery cells, 2. have overall the same as well as different battery types, and 3. have a variety of battery operation to examine transfer learning in different circumstances. Battery types are distinguished among others regarding materials and nominal capacity. Regarding battery operation charging, discharging, and storage of the batteries are relevant. This variety ensures the validity of the results for different battery types and operational load.

7.2 Design of Experiments

The experiments aim at answering "when" and "how to transfer" an SOH forecasting model from the source to the target domain. For developing the parametric transfer the essential model hyperparameters are the learning rate α and the number of frozen layers n_{frozen}. In addition, transfer learning is compared to the benchmarks proposed in Section 7.1.3. Further examinations are on the number of samples n_{samples} and their distribution over different batteries and aging states (Data split type). Model users are also interested in how the number of samples and data split type are influenced by the similarity of source and target data set.

The target data sets for training and validation have 128, 256, 512, 768, and 1024 samples. This is limited by the NASA random data set which has only 1183 samples. The learning rate is one of the common values of 0.001, 0.0001, and 0.0001. Training did not converge with higher learning rates. The RMSE is chosen as evaluation metric because it is common for regression problems. Furthermore, compared to the MSE, is has the same unit as the predicted output value which enables the direct comparison of the model's performance with a predefined error threshold. From the three measures by which learning might be improved due to a transfer introduced in Section 2.2.5, the lower asymptote is considered as the most important measure here because the final model performance after transfer learning is the most relevant. Thus, these experiments and their discussion focus on the final model performance.

Currently, no public battery aging data set from real-world vehicle operation exists. Thus, transfer learning for SOH forecasting with the laboratory data presented in Section 5.1 is examined to enhance comparability and reproducibility. The source model is obtained following the method presented in Chapter 6 using the DD data set as previously published [1]. The DD data set provides an ideal starting point for these experiments because of its similarity to CL. The source data set contains 3127 training samples and each 391 validation and test samples. Fine 2D stressor table, variant A is used because this showed the best result in Section 6.3 (Stressor table configuration: Tables 6.2 and 6.3). Further, it is assumed that the fine signal interval width enables the model to learn the target data better. Batteries in the Oxford and NASA random data set have a lifetime below 400 cycles. This prohibits w_w of 400 and 530 (W5 and W6) as well as the corresponding grouped data sets W10 to W12. Further, a combined w_W of $\{25, 50, 500\}$ cycles (W9) showed promising results regarding generalization in the source domain. Thus, W9 and $w_s = 25$ are selected. The hyperparameters, model complexity, and model performance of the source model on the source data are shown in Table E.2.

For better comparability, the benchmark MLP uses the same amount of layers and neurons as the source model. For TO and STM, only the learning rate 0.001, 0.0001, and 0.0001 and the batch size $bs \in \{16, 32, 64, 128\}$ are hyperparameters of a grid search. For SO, the source model is simply run in prediction mode as it is already trained only with the source data.

Compared to previous work [1] shown in Chapter 6, the min-max normalization is adapted to be suitable for the transfer learning. In previous work, min-max normalization was executed feature-wise using the source training data, i.e., each feature was normalized separately based on its minimum and maximum value in the source training data. The minima in the source and target data are mostly zero because the dwell time for every stressor type is zero. However, the maximum values of the target training data are different as stated in Section 6.1 ($P(X_\mathrm{S}) \neq P(X_\mathrm{T})$). Furthermore, determining the maxima of the features of the test target data is not possible at transfer time, because the test target data becomes only available in the future. Thus in this work, min-max normalization is executed in two groups of features: First, for the $SOH(t_1)$ and second for all other features, i.e., the stressor values from the stressor tables. The maximum of all stressor values of the source training data is used for normalization. Further, the same scaling from the source data for all target data ensures that the weight of the stressor values is the same for all features. In addition, this eases the model's transfer because the model does not need to adapt to a different normalization of the target data.

Experiment 1: How to transfer? Freezing & Learning rate

In general, n_frozen and α are expected to jointly influence the convergence speed and also the final target model performance. n_frozen measures how much information from the source domain is kept at transfer time. Thus, it is especially interesting as it also is a measure of similarity between source and target domain. Presumably, a frozen model will perform well, if the domains are similar. Contrarily, if less information on the future aging is given in the target domain training data, freezing is also beneficial to prevent negative transfers.

Experiment 2: When to transfer? Data splits & no. of samples
After examining the selection of the best model regarding n_{frozen} and α in Experiment 1, the question "when to transfer" concerns the suitability of the target data specified by the three data splits FO, MN, and ONM as well as the number of samples.

Experiment 2a: Closed-Loop
As visible in Table 5.1, the batteries in DD and CL have the same battery characteristics. Only the battery operation deviates regarding the charging protocols. Thus, DD and CL are similar domains, providing an ideal situation for transfer learning.

Experiment 2b: ISEA, NASA random, and Oxford
The transfer to the ISEA data set provides a change in cathode material from LFP to NMC and a smaller SOC range. Batteries in automotive applications like BEVs experience VC discharging in contrast to simple CC discharging of batteries in laboratory operation [2]. Compared to the other data sets, in the NASA random data set not CC discharging, but stepwise randomly changing discharge rates are applied. Not only cyclic aging, but also calendar aging is important in automotive applications for battery aging which is considered by the Oxford data set.

Because of these differences that influence the stressor data ($\subset P(X_{\text{T}})$) and the predictive function $f_{\text{T}}(\cdot)$, these three data sets are expected to be more challenging for successful transfer learning than the CL data set. Furthermore, the NASA random and Oxford data sets are a step towards applying the SOH forecasting model and its transfer to BEV operational aging data.

Each non-test target data set was randomly split for training and validation by the ratio of 90:10. The validation MSE was set as metric for early stopping [247] with patience of 5 epochs and minimum delta of 0 to avoid overfitting on the training target data. A maximum of 90 epochs was used. The constant hyperparameters from the source model training are not changed: Optimizer Adam, MSE as loss function, and a linear activation function of the output layer. The kernel regularizer was kept for the transfer to prevent overfitting. Version 2.8.0 of TensorFlow was used as back-end including version 2.8.0 of Keras.

7.3 Results

When selecting a model regarding the hyperparameters n_{frozen} and α for a given target data set, the validation and the test target RMSE are possible selection criteria. Both have a different distribution of the output values because of the temporal data split specified in Section 7.1.2 which leads to different aging rates for samples from the BOL compared to those from the EOL (see Figure 5.1). Due to these assumptions of data availability for the model selection at transfer time only the validation target set is available, even though overall test target RMSE is interesting because it indicates the model's performance when run in production.

Experiment 1: How to transfer? Freezing & Learning

Experiment 1 not only evaluates which n_{frozen} and α are optimal in the majority of scenarios, but also evaluates the reduction of model performance on the test data set due to the selection criteria only being the validation RMSE. Exemplary, Figure 7.3 shows the model's RMSE of all three α with different n_{frozen} for ISEA, MN with 128 samples. Selecting by the test RMSE (rectangles) $\alpha = 0.001$ and $n_{\text{frozen}} = 7$ would be optimal, but by the validation RMSE (crosses) $\alpha = 0.0001$ and $n_{\text{frozen}} = 0$ is best.

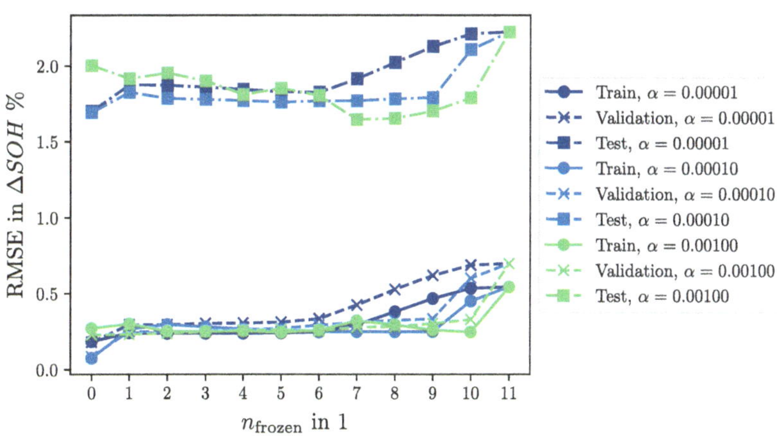

Figure 7.3 Exemplary model selection (ISEA, MN, 128 samples)

As depicted in Figure 7.4 for all four target data sets and selected by the validation RMSE, freezing no layer is the best choice in 95% of the data splits ($n_{\text{frozen}} = 0$). However, when selecting by the test RMSE this would only be the case in 45% of the data splits. 18% of all data splits have a better test RMSE with freezing the first layer instead of none. This indicates that in the majority of the data splits no general features have been learned from the source domain. Compared to the SotA procedure in the field of computer vision this is potentially due to the limited amount of source training data. $n_{\text{frozen}} = 10$ selected by the test RMSE is only optimal to prevent a negative transfer in the case of MN and CL as further discussed in Experiment 2.

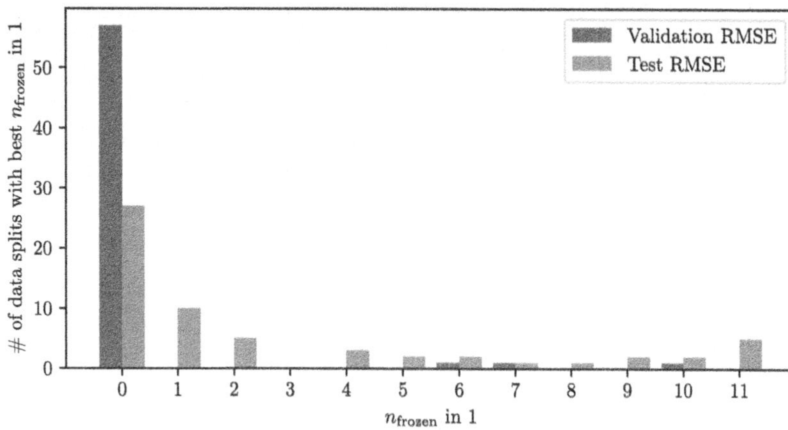

Figure 7.4 Best n_{frozen} selected by validation and test RMSE of all four target data sets

Further, for α the selection based on validation and target RMSE diverges even more as shown in Figure 7.5. For 72 % of the data splits 0.0001, which is also the source model's α, is the best α when selected by the validation RMSE, but for 60 % of the data splits 0.001 is the best when using the test RMSE. Overall, these discrepancies lead to an average deviation of the test RMSE when selecting n_{frozen} and α by the validation RMSE of 23 % with a SD of 28 %. Unfortunately, this error has to be accepted because the test RMSE itself is not accessibly for model selection. In the following, for each data split the best model is selected based on the validation RMSE.

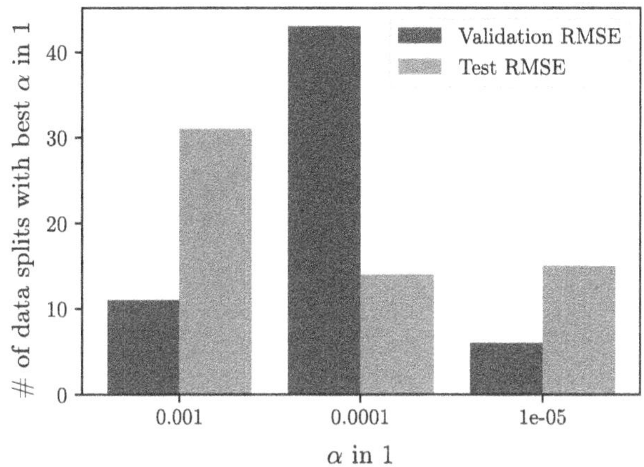

Figure 7.5 Best α selected by validation and test RMSE of all four target data sets

Experiment 2: When to transfer? Data splits & sample no.
Figure 7.6 depicts the performance of the best model selected by the validation RMSE on training, validation, and test data as well as the SO, TO, and STM benchmarks of all four target data sets. First, CL as target data set and then the remaining three target data sets are analyzed.

Experiment 2a: Closed-Loop
Figure 7.6a shows for the CL target data set FO that the training and validation RMSE are on the same level as the source RMSE (see Table E.2 W9 for comparison). The test RMSE is higher overall and just acceptable for an application of the model, e.g., by fleet managers who are interested in forecasting degradation from 100 % down to approximately 80 %. The test RMSE is decreasing once more data becomes available for training. This is coherent with the expectations. Independently of the amount of data only STM can compete with transfer learning. TO only reaches a similar model performance with 1024 training samples. For MN, a negative transfer is observed that leads to an RMSE of around 5 which is too high for a practical model application. This is caused by a lack of degradation information about the knee point. Even with 1024 samples not sufficient information about aging at EOL and the knee point is provided because 1024 samples only correspond to the first 250 cycles and a minimum SOH of 93 %, but the earliest knee point of CL is at around 500 cycles (see Figure 5.1). For MN, also the benchmarks SO and STM perform better than transfer learning and TO is as bad as transfer learning. This underlines

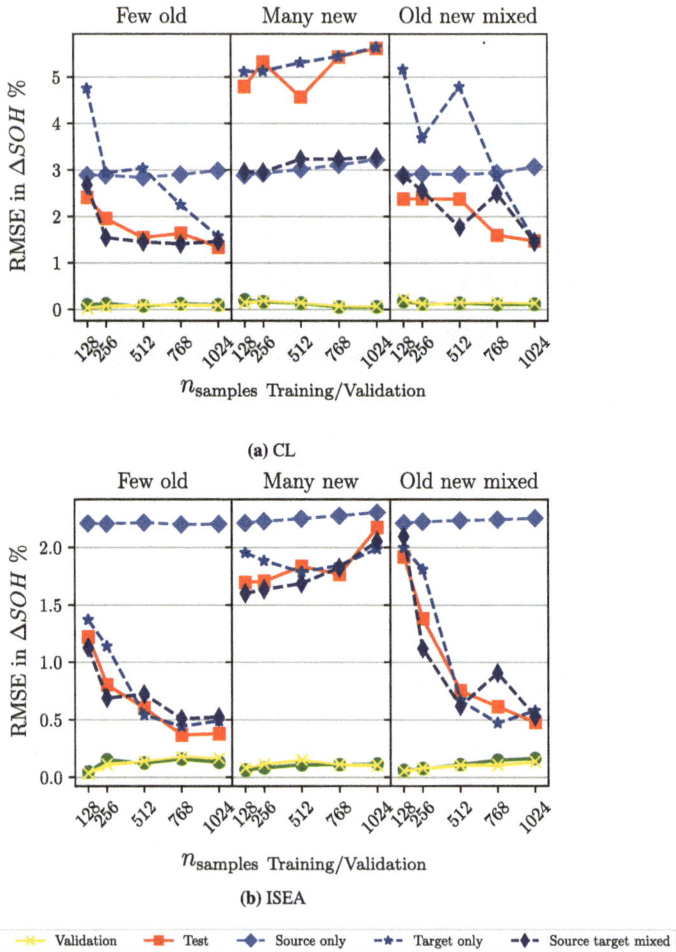

Figure 7.6 Best models selected by validation RMSE of all four target data sets and their benchmarks

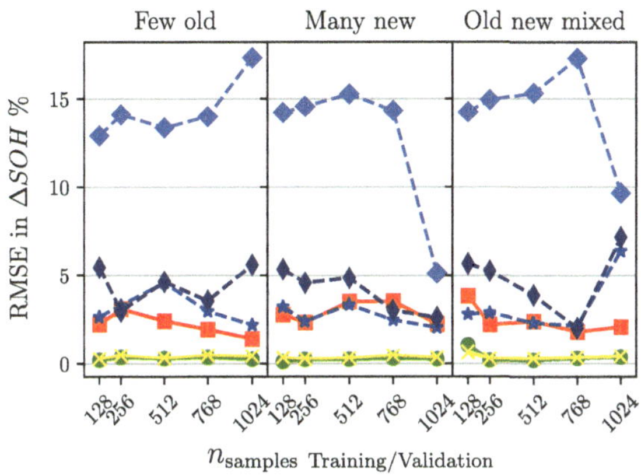

(c) NASA random

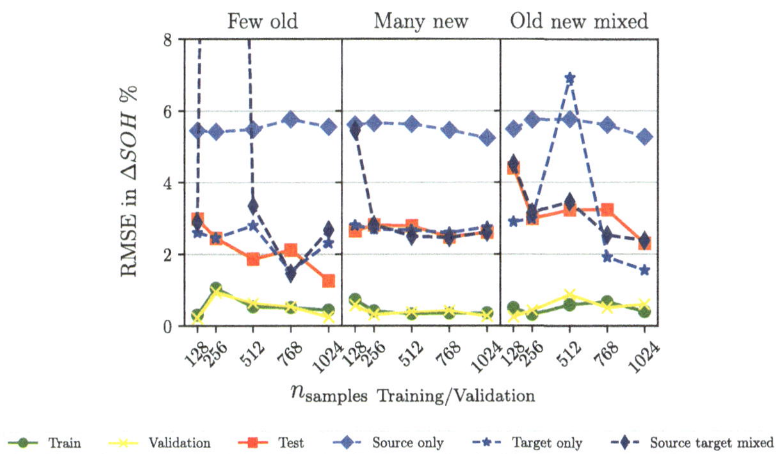

Figure 7.6 (Continued)

the unsuitability of the data of MN for transfer learning. For the ONM, the model performance is similar to FO. Thus, adding MN to FO does not provide additional relevant information to the model. Again TO performs worse than the transferred model.

Experiment 2b: ISEA, NASA random, and Oxford
The results using the target data sets ISEA, NASA random, and Oxford shown in Figures 7.6b to 7.6d respectively confirm that FO is the preferred scenario of data availability. Only for NASA random and Oxford the test RMSE of FO and MN is similar, but still FO is better. Presumably, this is caused by a constant aging rate of NASA random and Oxford (see Figure 5.1). Contrarily, DD, CL, and ISEA show superlinear degradation and, thus, have a knee point. This knee point makes forecasting with the MN data split more difficult than when having linear degradation because data of the accelerated aging after the knee point is missing. Consequently, the test RMSE of ONM of CL and ISEA is better than of MN because ONM contains information about the knee point.

Overall, the SO benchmark is worse than the other benchmarks and transfer learning except in 16 % of the cases. This indicates that any data from the target data set improves the model's performance. The comparison of the TO and STM benchmarks with transfer learning leads to no unambiguous results as each of the three is performing slightly better in some cases. Unfortunately, no clear pattern was identifiable under which circumstances which of the three is preferable. Probably, a more extensive source data set is required for a clear added value of transfer learning. One could for example use four of the presented data sets as source data set and only one as the target data set.

7.4 Summary and Contribution to the Research Questions

This chapter showed under which conditions transfer learning enables SOH forecasting based on a comprehensive study using four known public target battery data sets (RQ.2): Having data of few aged batteries shall be preferred over data of many young batteries, especially in the case of superlinear degradation with knee points. In the case of data of few aged batteries, having data of more aged batteries from the target domain will improve model performance. This corresponds to the results of [453, p. 9] who point out that at least one SOH trajectory of an aged battery should be known, if only short-term degradation data is available of the other batteries. Also, the problem of an available metric (Validation vs. test RMSE) was discussed which prohibits selecting the best possible model. In contrast to SotA transfer learning in computer vision, freezing no layers was preferable for the transfer in 95 % of the experimental scenarios. This is due to the limited amount of source data and coherent to the results of [326].

As the presented method in Section 6.1 used in Chapter 6 and 7 is purely data-driven it may only provide good interpolations within the range of the training data. The data sets used in Chapter 6 and 7 originate only from laboratory operation of battery cells. In particular for Chapter 6, the DD battery data was obtained from batteries cycles with 100% ΔSOC. Cycles with smaller ΔSOC were neither part of the data basis, nor were they used for validation. Furthermore, discharging was with CC over the entire data set. Of the data used in Chapter 7, only the NASA random and Oxford data sets come close to BEV operational data with VC discharge and calendar aging respectively. So the findings from Chapter 6 and 7 are limited for VC discharging like in automotive applications. Thus, the next Chapter 8 applies a data set of battery systems from a BEV fleet and examines if the features are capable of capturing VC discharging for NI cycles (NI-VC).

8 Battery System State of Health Forecasting

In Section 8.1, the method from Section 6.1 is adapted and extended to the battery data from real-world fleet operation presented in Section 5.2. Section 8.2 presents the DOE which is followed by the evaluation of the defined experiments in Section 8.3. The two types of stressor data defined in Section 8.1.2 are first evaluated separately and then compared with each other. Finally, Section 8.4 applies one SOH forecasting model in two scenarios under changed operational conditions as proposed in Section 3.3.

In this chapter, SOH means SOH_E and not like in the other chapters SOH_C. The method and results that are presented throughout this chapter are primarily based on previous publications [3; 9; 10].

8.1 Method

Building upon the method from Chapter 6, this chapter's method is introduced in Section 8.1.1. Section 8.1.2 presents features to encode battery operation, in particular accessible feature as alternative to the histograms used previously in Chapters 6 and 7.

Supplementary Information The online version contains supplementary material available at https://doi.org/10.1007/978-3-658-43188-4_8.

F. von Bülow, *A Data-Driven Fleet Service: State of Health Forecasting of Lithium-Ion Batteries*, AutoUni – Schriftenreihe 170,
https://doi.org/10.1007/978-3-658-43188-4_8

8.1.1 State of Health Forecasting

For this chapter, the base method from Chapter 6 depicted in Figure 6.1 is applied for several reasons: First, it fulfills the key criteria examined in the structured literature review in Chapter 4. Second, the histogram features presented are suitable to encode the variability of real-world operation, especially the VC signal which is NI over multiple cycles (NI-VC) as analyzed in Section 2.1.6.2. Third, the window-based method allows easy adaption to other features for different user groups as argued in Section 8.1.2.

Since in the BEV operational data more hold, i.e., parking phases, are apparent, the choice of the cycle number for the window definition needs reconsideration: The window $[t_1, t_2]$ was defined by the cycles in Chapters 6 and 7. However, for better comparability among cycles with different ΔSOE it is favorable either to use the EFC or the time as battery aging in real-world applications is a combination of cyclic and calendar aging. For better comparability with other methods, which are normally based on the EFC, a sliding window approach based on the EFC is chosen to derive training samples.

8.1.2 Stressor Data

In this chapter, different stressor data are extracted in step one in Figure 6.1 that are input of the ML model: As in Chapter 6, histograms are used, but also other, more accessible stressors for non-domain experts inspired by existing work on SOH forecasting are motivated. Thus, by using different stressor data features the trade-off between model performance and feature interpretability for the model's user is examined. Therefore, the same time windows, i.e., the same $SOH(t_1)$ as feature and ΔSOH as output variable are kept. This maintains the output value distribution and, thus, improves comparability among the different stressor data

Features from the frequency domain, e.g., by a wavelet transformation proposed in Section 6.1.1, are not used because the sampling time of 10 s of the BEV data set presented in Section 5.2 is too low to obtain meaningful information of the frequency of the time series.

Histograms

Histograms of different dimensionalities (1D, 2D, 3D) based on C-rate, temperature, and SOE have been used as stressor data in Chapter 6. Therefore, the signal interval width is defining the sampling granularity of the signals' bins. E.g., the temperature signal could be split into intervals of 1 °C or 2 °C.

Accessible Features

The more accessible features described in Table 8.1 are either inspired by other papers on SOH forecasting or motivated by possible user groups of SOH forecasting models. They either reflect cyclic or calendar aging.

The term "accessibility" here refers to features that are easy to use and understand by the user groups. For example, users with little BEV experience can interpret 1000 km of driving as with an internal combustion engine (ICE). They may also have a gut feeling for the usual number of their AC- and DC-charging events per week why this is chosen as an accessible feature. Also, as values based on experience for low or high fuel consumption exists, such experience for the energy consumption, e.g., in kWh/100 km, may develop. From there an estimate of the energy throughput E_{thrpt} required for a certain distance and terrain can be derived.

Private BEV users may choose a certain driving mode to adapt the vehicle's performance, e.g., regarding regenerative braking strength, or decreasing the accelerator pedal responsiveness for quick acceleration [454]. Further, an ECO mode could restrict the maximum power output to 80 %, i.e., the maximum discharge current $I_{\mathrm{dr,max}}$ assuming CV. The maximum available discharge current is easily adaptable and a method to slow down battery aging.

For the operation and management of BEV fleets different viewpoints are relevant [6; 7]: In the short term, a role like a shift or operation manager might consider the amount of available BEVs and their SOE at shift beginning ($SOE_{\mathrm{park,mean}}$). Further, a head of operations might be responsible for vehicle replacements by a new vehicle generation. Both shall not be required to acquire deep technical understanding, just as a private BEV user. However, they may be supported by a vehicle data analyst who has the technical understanding.

Table 8.1 Overview of accessible features

	Name	Symbol	Unit	Reference
Cyclic aging	Cycles from t_1 until t_2	k	EFC	[239; 329; 388; 389]
	Delta mileage	Δd	km	[301; 455]
	Number of AC charging processes	n_{AC}	1	[393]
	Number of DC charging processes	n_{DC}	1	[393]
	AC charging energy throughput	$E_{thrpt,AC}$	Wh	
	DC charging energy throughput	$E_{thrpt,DC}$	Wh	
	Absolute max. charging current	$I_{ch,max}$	C	[301; 391; 402]*
	Mean charging current	$I_{ch,mean}$	C	[301; 391; 402]*
	SD of charging current	$I_{ch,SD}$	C	[301; 391; 402]*
	Absolute max. current while driving†	$I_{dr,max}$	C	[301; 391; 402]*
	Mean current while driving	$I_{dr,mean}$	C	[301; 391; 402]*
	SD of current while driving	$I_{dr,SD}$	C	[301; 391; 402]*
	Min. temperature	T_{min}	°C	[391]
	Max. temperature	T_{max}	°C	[391]
	Max. SOE	SOE_{max}	%	
	SOE lift‡	ΔSOE	%	
Calendar aging	Parking time	Δt_{park}	h	[229]
	Mean parking SOE	$SOE_{park,mean}$	%	[229]
	Mean parking temperature	$T_{park,mean}$	°C	[229]
	Time from t_1 until t_2	Δt_{total}	h	[391]

* [402] use statistical properties of I_{disch}. [391] use I_{mean}. As it remains unclear whether it refers to charging or discharging, $I_{ch,mean}$ and $I_{dr,mean}$ when $I \neq 0$ A are derived. Also, the used data set does not include Q_{thrpt} and SOC. Thus, the correlating E_{thrpt} and SOE respectively are used instead. [301] propose distributional information of the C-rate during cycling. † NaN values are not included. $I_{dr,min}$ means $\min(I_{dr})$ if drive mode, i.e., including internal charging due to brake energy recuperation. Also, $I_{dr} < 0$ A for discharging. ‡ It is assumed that $\Delta SOE = SOE_{max} - SOE_{min}$. In their data set, [391] only have a ΔSOE of 60 % and 100 %.

8.2 Design of Experiments

Analogous to the defined two types of stressor data in Section 8.1.2, histogram-based features and accessible features, two series of experiments are run as presented in Sections 8.2.1 and 8.2.2 respectively.

8.2.1 Histogram-based Features

For showing the suitability of the chosen SOH forecasting method on a data set from battery systems in BEV fleet operation, it is essential to find good hyperparameter values of the stressor extraction described in Sections 6.1.1 and 8.1.1. The five hyperparameters are examined in decreasing order of the influence of the parameter on the model performance based on experience from this chapter and Chapter 6: First, experiment 1-3 examine w_{W} and w_{S} of the time window as well as the signal interval width for current, temperature, and SOE. Thereby, the right balance of input features and training samples shall be found. Further, experiment 4 and 5 examine different combinations of the current, temperature, and SOE signals to stressor tables for charging, discharging, and parking mode. Also, the capability of the SOH forecasting model to generalize is investigated.

At the start of the DOE, the initial parameters are set as follows: $w_{\mathrm{S}} = 5$ cycles, except if $w_{\mathrm{W}} = 1$ cycle as $w_{\mathrm{S}} \leq w_{\mathrm{W}}$ should be true. The signal interval width is coarse with 2D histograms and BEV mode A for the 2D stressor tables based on Chapter 6. The 2D stressor tables, variant A were modified by substituting "discharging" by "driving" which now includes micro-cycling due to regenerative braking.

Experiment 1: Window width (w_{W})

First, w_{W} determines the forecast horizon and the number of cycles corresponding to the histograms aggregating the operational data. Thus, it also influences the distribution of the output value ΔSOH which the ML model has to learn. The selection of w_{W} depends on how many cycles can be sensibly aggregated together to offer sufficient variance among the histograms for the ML model to distinguish the samples. The selection of w_{W} is inspired by forecast horizons listed in Table 4.3 like 30 cycles [417], 40 cycles [236], and 25 cycles to 530 cycles [1] also presented in Chapter 6, but also 1 cycle [346; 389; 391; 392; 402; 401]. Furthermore, the relevance of the forecast horizon for BEV users and fleet managers is considered. Given the fleet's SOH trajectory in Figure 5.2 with up to 450 cycles, larger w_{W}, like in Chapter 6 with data up to around 1000 cycles, prohibit generating at least a few samples from each battery system. Thus, partially different w_{W} compared to Chapter 6 are chosen here: $w_{\text{w,single}} \in \{1, 5, 10, 25, 50, 100\}$ (W1 to W6). These w_{w} are also combined because this showed improved generalization on battery cell data in Chapter 6: $w_{\text{w,combined}} \in \{\{10, 25\}, \{25, 50\}, \{50, 100\}, \{10, 25, 50\}, \{25, 50, 100\}\}$ (W7 to W11).

Experiment 2: Window shift (w_s)
Second, w_{s} influences how many cycles the window is shifted during sample generation, i.e., how many samples are generated. Thus, its setting needs to ensure sufficient sampling points for the ML model to learn and to acquire the ability of good interpolation in between the sampling points, i.e., a good model fit. w_{s} is set depending on w_{w} because of their interaction to $w_{\text{s}} \in \{5, 10, 25, 50, 100\}$ cycles with $w_{\text{s}} \leq w_{\text{w}}$ to avoid skipping operational data, as mentioned before.

Experiment 3: Signal Interval Width
Third, the signal interval width, i.e., the width of the histograms' bins, determines the amount of features in the histograms. A finer signal interval width enables the histograms to encode the underlying time series data more precisely. Table 8.2 specifies fine, medium, and coarse signal interval width (F, M, and C respectively) for current, temperature, and SOE. Combining different signal interval widths like different w_{W} is not possible because they have different input feature shapes requiring different input layers. Compared to Chapter 6 with the data presented in Section 5.1.1.1, the granularity of the current and temperature bins is adapted because their overall range is different in this data set presented in Section 5.2. For the C-rate an additional interval for zero current is introduced, representing the parking mode.

Table 8.2 Signal interval width for current, temperature and SOE

		C-rate in C	T in °C	SOE in %
Bin range		[−1.5, 1.5]	[−20, 50]	[0, 100]
Bin width	Fine (F)	0.1	3	5
	Medium (M)	0.3	5	10
	Coarse (C)	0.5	10	20

Experiment 4: Histogram Dimension
Forth, the dimension of the histogram can either be 1D, 2D, or 3D, i.e., one, two, or all three signals of T, I, and SOE are combined in one histogram. A lower dimensional histogram can be seen as a column-wise aggregation of a higher di-

mensional histogram. Thus, lower dimensional histograms have fewer features, but lack a representation of the signals' interdependence.

Experiment 5: Separation of BEV Modes
Fifth, histograms depending on the BEV mode are examined as listed in Table 8.3: Separate histograms for charging, driving, & parking (BEV mode A), AC-/DC-charging, driving, & parking (BEV mode B), and joint histograms for all BEV modes (BEV mode C) are differentiated. Thus, BEV mode C has only three histograms ($T\&SOE$, $I\&SOE$, $I\&T$) and does not differentiate the BEV mode.

Table 8.3 Combined signals for 2D stressor tables by BEV mode

	$T\&SOE$	$I\&SOE$	$I\&T$
BEV mode A*	Charging, Driving, Parking	Charging, Driving	Charging, Driving
BEV mode B	AC-Charging, DC-Charging, Driving, Parking	AC-Charging, DC-Charging, Driving	AC-Charging, DC-Charging, Driving
BEV mode C	No differentiation	No differentiation	No differentiation

* Called variant A in the previous Chapter 6.

8.2.2 Accessible Features

These experiments aim at finding a set of accessible features for non-domain experts, i.e., of non-histogram scalars that represent the battery load as motivated in Section 8.1.2. In experiment 1, only $SOH(t_1)$ is used to determine suitable w_s for $w_w \in \{10, 25, 50, 100\}$ from experiment 1 with histogram-based features. Then, in experiment 2a and 2b cyclic and calendar aging features are analyzed respectively. Next, all of these features are combined in experiment 3.1 and only the best of experiment 2a and 2b are combined in experiment 3.2. Table 8.4 summarizes all experiments with accessible features.

8.2.3 General Settings

Each data set generated with a different setting of histogram-based or accessible features, as described in Sections 8.2.1 and 8.2.2 respectively, was split to enable evaluation of the model's generalization performance. First, to enable evaluation of the model's generalization performance on different vehicles than in the training

Table 8.4 Other feature sets used as stressor data

		Start	Cyclic Aging							Calendar Aging			
		$SOH(t_1)$	k	Δd	n_{AC} & n_{DC}	$E_{thrpt,AC}$ & $E_{thrpt,DC}$	$I_{ch,dr}$*	T_{min}, T_{max}	SOE_{max}, ΔSOE	Δt_{park}	$SOE_{park,mean}$	$T_{park,mean}$	Δt_{total}
Experiment No.	1	x											
	2a.1	x	x										
	2a.2	x		x									
	2a.3	x			x								
	2a.4	x				x							
	2a.5	x					x						
	2a.6	x						x					
	2a.7	x							x				
	2b.1	x								x			
	2b.2	x									x		
	2b.3	x										x	
	2b.4	x											x
	3.1	x	x	x	x	x	x	x	x	x	x	x	x
	3.2	x				x	x						

* $I_{ch,dr} = \{I_{ch,max}, I_{ch,mean}, I_{ch,SD}, I_{dr,min}, I_{dr,mean}, I_{dr,SD}\}$ with the BEV modes "ch" as charging and "dr" as driving.

and validation data set, 5 % of the vehicles were chosen as test vehicles (2. Test). I.e., this tells about the model's forecast performance on new vehicles that recently joined the fleet without any time series used for training sample generation. Second, the samples of the remaining vehicles were split into the training, validation, and test data (1. Test) by the ratio of 80:10:10, reaching an overall ratio of 76:9.5:9.5:5 for training, validation, and the two test sets. The training data set was used to fit the model's parameter and the validation data set served for early stopping and hyperparameter optimization. The 1st test data set enables evaluation of the model's generalization performance on the same vehicles as in the training and validation data set. E.g., neighboring time windows may not be in the same data set.

Min-max-normalization is applied to each data set based on the minimum and maximum values of the training data set. Similar to Section 7.2, $SOH(t_1)$ is scaled separately from the other features. The histograms are also scaled separately for the BEV modes, i.e., AC-/DC charging, discharging, parking, to preserve the relations within each BEV mode. Otherwise the values of charging and discharging would be very small due to larger parking times. The accessible features are scaled separately from each other because they have limited physical relation to each other.

As ML model, an MLP is chosen because it showed good results in comparison to GPR as shown in Section 6.3 as well as over other ML and learning (DL) models as another previous publication shows [4]. For each data set a corresponding hyperparameter optimization was executed using Hyperopt version 0.2.7 with TPE which resulted in a final model for each data set [267; 268]. Hyperopt parameters were set to a maximum of 500 evaluations and 100 random startup evaluations. The MSE on the validation data set was set as optimization metric for the hyperparameter optimization. The variable hyperparameters of the MLP and their range are chosen form experience and are shown in Table 8.5. The constant hyperparameters of the MLP are the optimizer Adam, 300 epochs, MSE as loss function, ReLU activation function of the hidden layers, and a linear activation function of the output layer. Early stopping [247] is applied so that training stopped earlier than 300 epochs if the validation error did not decrease for more than 10 epochs. Version 2.8.0 of TensorFlow is used as back-end including version 2.8.0 of Keras.

Table 8.5 Overview of the used hyperparameters of the MLP

Hyperparameter		Values
Batch size		32, 64, 128
Learning rate α		log-uniform distribution [0.0001, 0.01]
Number of layers		uniform distribution [1, 15]
Neurons per layer		log-uniform distribution [20, 500]
Choose one, others are 0:	Regularization parameter λ_1 or λ_2 [247; 249; 256]	log-uniform distribution [0.001, 0.1]
	Dropout rate [449]	log-uniform distribution [0.1, 0.5]

Compared to the experiments in Chapter 6, the granularity of the current and temperature bins is adapted because their overall range is different in this data set. For the C-rate an additional interval for zero current is introduced, representing the parking mode.

8.3 Results

First, the results of the models trained with histogram-based features are evaluated in Section 8.3.1 and then those trained with the accessible features in Section 8.3.2. Section 8.3.2 concludes with a comparison of both feature types.
The hyperparameter optimization of the MLP never converges choosing the maximum of the value range for the number of layers and neurons per layer in Table 8.5. This underlines a suitable choice of the value range of the hyperparameters.

For model comparison trained on data sets with different w_{w}, R^2 is used instead of the RMSE because w_{w} changes the output value distribution (ΔSOH) and R^2 normalizes the error by the variance of the output value distribution.

Two criteria are defined for the model evaluation: First, the SD σ of the model's performance on the data sets $s \in S = \{training, validation, test_1, test_2\}$ shall be small to avoid overfitting:[1]

[1] With $\sigma_{R_S^2} = \sqrt{\frac{1}{|S|} \sum_{s \in S} \left(R_s^2 - \mu_{R_S^2}\right)^2}$ and $\sigma_{RMSE_S} = \sqrt{\frac{1}{|S|} \sum_{s \in S} \left(RMSE_s - \mu_{RMSE_S}\right)^2}$ as well as the corresponding $\mu_{R_S^2} = \frac{1}{|S|} \sum_{s \in S} R_s^2$ and $\mu_{RMSE_S} = \frac{1}{|S|} \sum_{s \in S} RMSE_s$ respectively where R_s^2 and $RMSE_s$ are R^2 and RMSE on the respective data set s.

$$\sigma = \sigma_{R_S^2} < 0.1 \text{ or } \sigma = \sigma_{RMSE_S} < 0.1\,\%SOH. \tag{8.1}$$

Second, if fulfilling the first criterion the mean μ of the model's performance on training, validation, and both test data sets shall be minimal:

$$\min \mu_{R_S^2} \text{ or } \min \mu_{RMSE_S}. \tag{8.2}$$

8.3.1 Histogram-based Features

The histogram-based features experiments are executed to analyze the effect of the following parameters on the model performance: w_W, w_S, signal interval width, histogram dimension, and separate histograms for each BEV mode.

Experiment 1: Window width (w_W)
The results of the models with a single window width are visualized in Figure 8.1. For all plots, the dotted horizontal line in the lower plot indicates the upper limit for the SD of the R^2 from the first evaluation criterion.

For models trained on a data set with a small w_W like 1 EFC and 5 EFC, the first criterion is fulfilled ($\sigma_{R_S^2} < 0.1$), but the mean performance $\mu_{R_S^2}$ is low with only 0.045 and 0.172 respectively. This is likely caused by the bigger resolution of the SOH signal, 0.36 % in Table 5.2, compared to the small ΔSOH during few EFCs, 0.0173 % per EFC and vehicle of the whole fleet. This confirms the problem of small ΔSOH during few EFCs that also appeared with laboratory battery cell data in Chapter 6. Thus, small w_W are not suitable due to low model performance.

For larger w_W like 25 EFC, 50 EFC, and 100 EFC, overfitting is observed as $\sigma_{R_S^2}$ increases gradually which is especially caused by a decreasing R^2 on the 2nd test data set, while R^2 of the training, validation, and 1st test data only move apart a little. This means for larger w_W that the ability to predict on new vehicles in the fleet is smaller than for vehicles whose data was used during training. By adjusting w_S in the next experiment the overfitting on the 2nd test data set shall be addressed. Overall, larger w_W show an increased performance on their training, validation, and 1st test data than smaller w_W.

In Chapter 6 with laboratory battery cell data, combining different w_W improved model and generalization performance compared to the models with the corresponding single w_W. The same comparison is made in Table 8.6 for vehicle battery system data: A reduction of SD over all data sets is observed, but also a decrease of R^2. For example, for $w_W = \{25, 50, 100\}$ SD decreases by 29.3 %, but also $\mu_{R_S^2}$

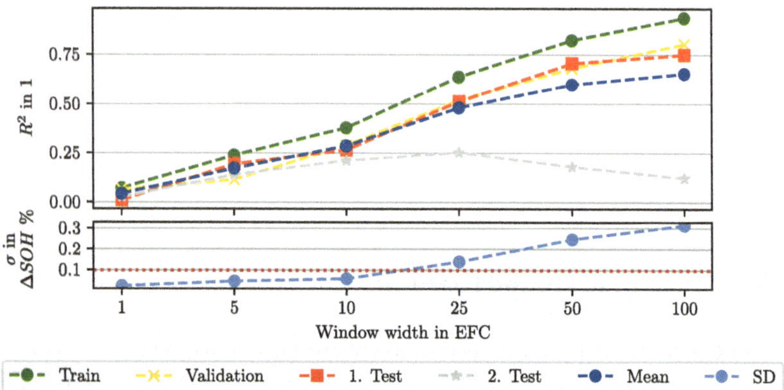

Figure 8.1 Model performance with different window widths w_w and histogram-based features

Table 8.6 Model performance with different w_w and histogram-based features

Window width (w_w)	σ_{R^2}	μ_{R^2}
25	0.1406	0.4807
50	0.2468	0.5986
100	0.3139	0.6547
{10, 25}	0.132	0.4267
{25, 50}	0.2269	0.5721
{50, 100}	0.3483	0.6061
{10, 25, 50}	0.2361	0.536
{25, 50, 100}	0.2218	0.607

decreases by 7.2 %. Presumably, due to the higher variability in the data from real-world vehicle operation compared to laboratory operation, models with single w_w are already less prone to overfitting. So there is no need to add further variability by combining w_w as it was for data from laboratory operation in Chapter 6. As model performance did not improve for combined w_w, these are not considered further, but experiment 2 aimed at improving generalization performance by adapting w_s.

Experiment 2: Window shift (w_s)

For the better performing $w_w \in \{10, 25, 50, 100\}$ from experiment 1, w_s was varied with the results displayed in Figure 8.2. The dotted horizontal line again visualizes the limit for SD defined in the first evaluation criterion. In this experiment for constant w_W, varying w_S are compared in each subplot of Figure 8.2 so the output value distribution does not change. Thus, the RMSE is used as performance metric. Comparing results from different w_W, i.e., different subplots of Figure 8.2, is only possible with limitations considering the different forecast horizons.

For all w_W in Figure 8.2 there exists an optimal w_S minimizing the SD of the RMSE on the training, validation, and both test data sets (1st evaluation criterion). These optimal combinations of w_W and w_S are chosen for the following experiments: $(w_W, w_S) \in \{(10, 5), (25, 10), (50, 25), (100, 25)\}$. These pairs fulfill the 1st evaluation criterion and are best regarding the 2nd evaluation criterion in their group.

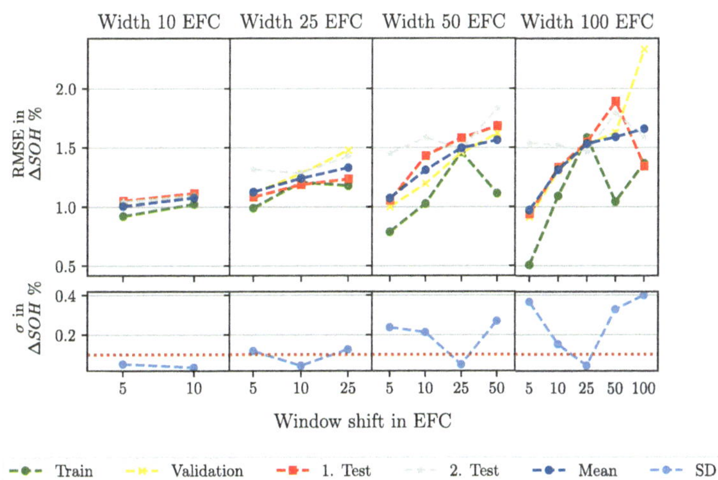

Figure 8.2 Model performance with different w_S and histogram-based features

The DOE in Section 8.2.1 stated that w_S impacts the amount of training samples and the difference between training samples due to overlapping time ranges of two neighboring windows. To analyze the second aspect in more detail, the feature disparity is introduced: The feature disparity is the mean difference of each histogram feature value between neighboring windows of the same vehicle in the data set. I.e., the mean is calculated over all histogram features, all pairs of neighboring windows, and all vehicles. The feature disparity is zero for the same window, i.e., the same histogram feature values ($w_S = 0$), close to zero for highly similar windows, and increasing the more dissimilar the histogram feature values are. As visualized in Figure 8.3, for increasing w_s the feature disparity increases independently of w_w because time windows overlap less.

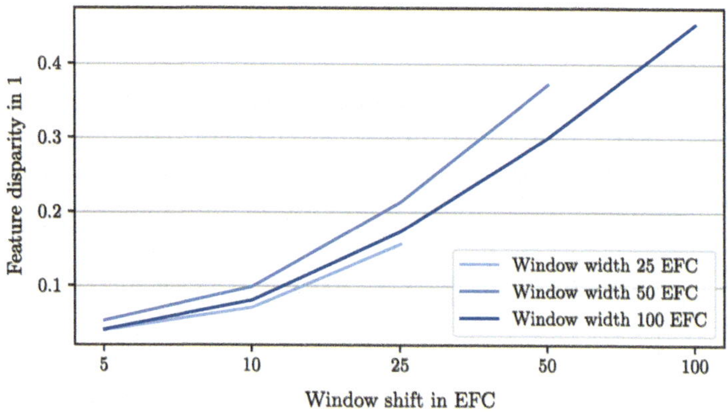

Figure 8.3 Histogram feature disparity over window shift w_s for several window width w_w

To analyze the potential dependence of the feature disparity on the model performance and generalization, Figure 8.4 depicts the model performance over the histogram feature disparity for all w_s and w_w run in experiment 2. At low feature disparities between 0.0 and 0.1 there is a high variance of R^2 reaching a minimum around a feature disparity of 0.2 and increasing for higher values. The mean of R^2 increases from a feature disparity of 0.0 until reaching a plateau at around 0.1. In conclusion, a feature disparity of around 0.2 offers a good trade off in-between model performance and generalization on training, validation, and both test data sets. Especially, the performance on the 2nd test data set also reaches its maximum at a feature disparity of around 0.2. This also means that w_s is an effective means to control overfitting of the models. This holds especially true for the 2nd test data set which was sometimes not well-fit in experiment 1. With a feature disparity of around 0.2 as reference point for new data sets computationally expensive training and optimization of multiple models may be reduced. Nevertheless, this hypotheses needs to be further evaluated in future experiments.

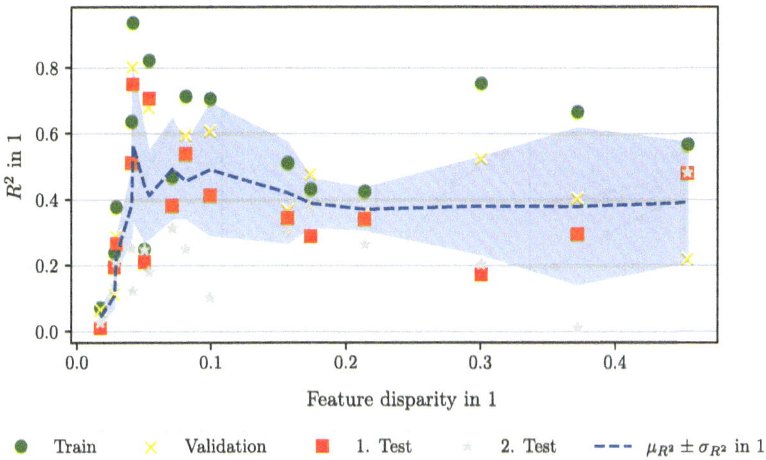

Figure 8.4 Model performance over histogram feature disparity for all window shift w_s and window width w_w run in experiment 2

Experiment 3: Signal Interval Width

As shown in Figure 8.5, a finer signal interval width increases the SD for most models because only the RMSE on the training data improves. This is indicating an increasing overfitting. With a finer signal interval width the number of features increases from 369 for coarse, to 958 for medium, and 3760 for fine but the number of samples is constant. This is problematic because it raises the curse of dimensionality: An exponentially increasing number of training samples is required to cover an increasing number of features and their combinations [256]. Furthermore, the ML models have +10 % parameters for medium and +249 % for fine compared to coarse signal interval width which is not dampened by regularization so the models overfit. The following experiments are continued with the coarse signal interval width because only the coarse signal interval width satisfies the 1st evaluation criterion for all w_w.

Experiment 4: Histogram Dimension

Similar to experiment 3, when increasing the histogram dimensions in this experiment, the number of features increases from 67 for 1D up to 369 for 2D and 1153 for 3D histograms. In Figure 8.6 the corresponding results are visualized. For w_w set to 10, 25, and 50 cycles, the 1st evaluation criterion regarding SD is fulfilled independently of the histogram dimension. However, for w_w = 100 cycles the 1st

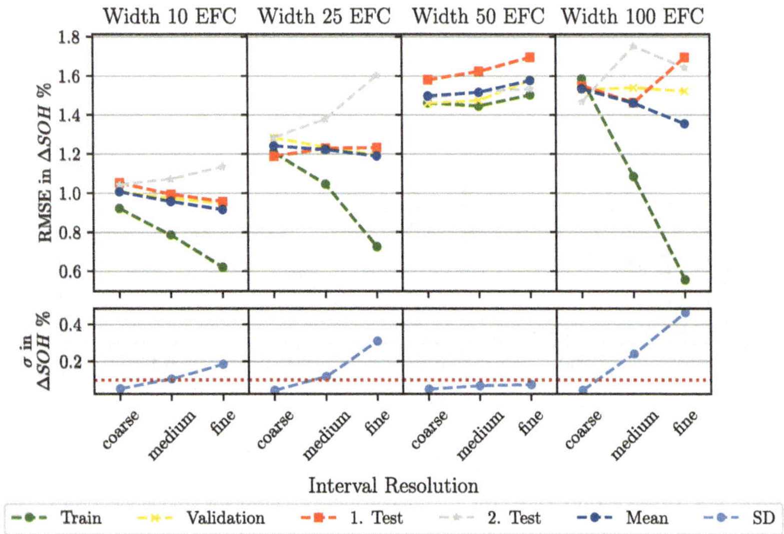

Figure 8.5 Model performance with different signal interval widths and histogram features

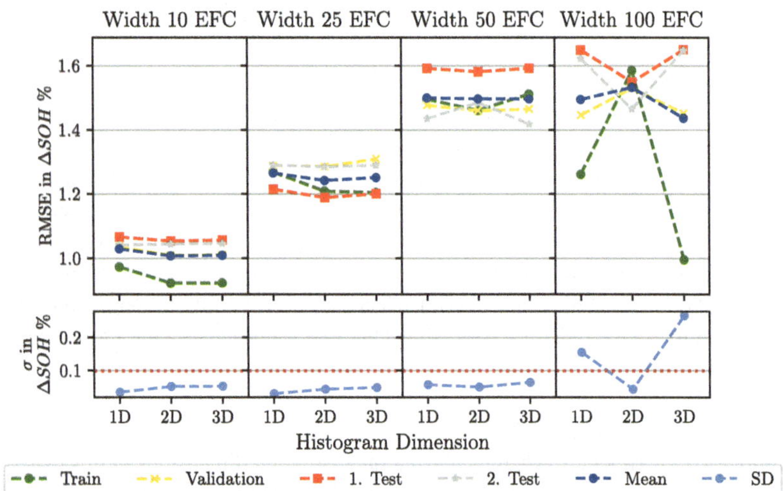

Figure 8.6 Model performance with different histogram dimensions

evaluation criterion is only fulfilled for 2D histograms and for $w_W = 50$ cycles SD is still lower for 2D histograms than for 1D and 3D histograms. For $w_W \in \{10, 25, 50\}$ cycles, the mean RMSE on training, validation, and both test data sets improves by 0.016 % SOH when switching from 1D to 2D and decreases by 0.0036 % SOH when switching from 2D to 3D. So the differences for these w_W caused by the histogram dimension are minor. This is not the case for $w_W = 100$ cycles where the model overfits on the training data for 1D and 3D, similar to the medium and fine signal interval width in experiment 3.

In general, the 2D histograms seem to provide a good balance between a sufficient representation of battery aging causes and low number of features. Besides, editing 2D histograms for virtual predictions, i.e., what-if scenarios, is easier and more conceivable as 3D histograms which are difficult to represent visually.

Experiment 5: Separation of BEV Modes

The separation of histograms by the BEV mode barley influences the model performance for $w_W \in \{10, 25, 50\}$ cycles as the results in Figure 8.7 show. As in the previous experiments, for $w_W = 100$ cycles the performance is more variable: When separating charging into AC- and DC-charging (B) compared to no separation at all (A) the SD increases by 219 % and the mean RMSE decreases by 1.4 %. The small mean RMSE improvement may be explained by only little information added by separating the charging modes. Already the current alone offers enough potential for differentiation of AC- and DC-charging in the histograms. In particular for the final phase at the end of DC-charging, especially the CV phase, when the current is controlled and as low as for AC-charging separate histograms for AC- and DC-charging (B) offer additional information.

Without BEV mode differentiation (C) the SD increases by 95 % and the mean RMSE increases by 1 % compared to A. Concluding, separate histograms by the BEV modes charging, driving, parking (A) are preferable and this variant is selected for the final models. Furthermore, this allows explicit distinction of different BEV modes during the model application.

The final models selected after experiment 5 and their hyperparameters are summarized in Table F.1 (accessible in the Electronic Supplementary Material).

Exemplarily, for the model trained on $w_W = 25$ and $w_S = 10$, Figure 8.8 shows the $\Delta SOH_{\text{predicted}}$ over the $\Delta SOH_{\text{measured}}$. A qualitative comparison of these plots with the same method applied on battery cell data from laboratory operation in Figure 6.6 shows that SOH forecasting with the given real-world battery system data performs worse. These results are expected to become better with finer time series data sampling and more precise SOH labels.

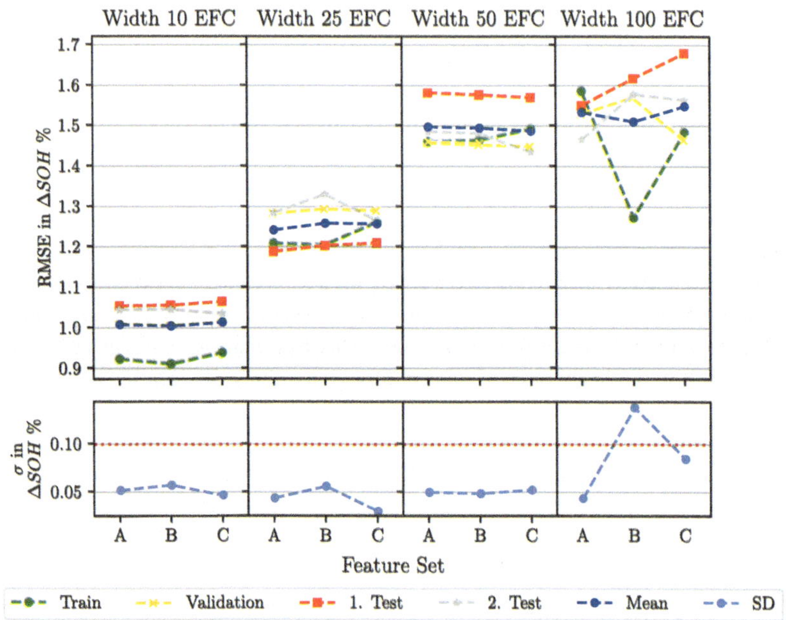

Figure 8.7 Model performance considering BEV-modes separately

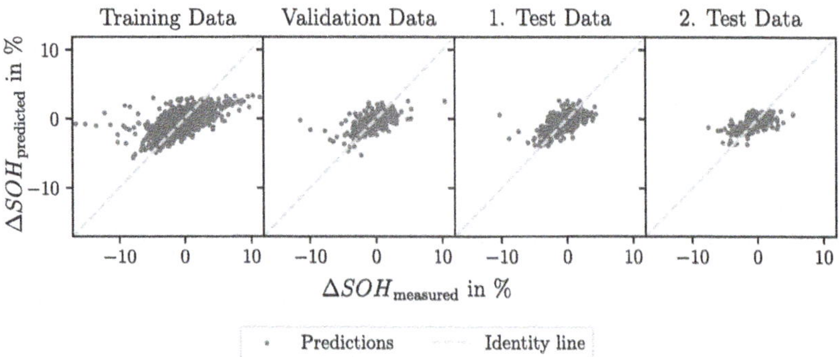

Figure 8.8 Model performance after last experiment, model H2 ($w_w = 25$ and $w_s = 10$), $\Delta SOH_{predicted}$ over the $\Delta SOH_{measured}$

The same predictions as in Figure 8.8 are displayed as moving boxplot compared to the mean measured SOH in Figure 8.9. Also, the 50 % and 90 % confidence zone for the prediction error are visualized which represent the upper and lower boundaries of the 50 % and 90 % best predictions respectively compared to the mean SOH curve. Until around 300 EFC the 50 % confidence zone has a width of around 1 % SOH. This means half of the predictions deviate less than 0.5 % SOH from the measured SOH. The 90 % confidence zone is significantly wider with around 3.5 % SOH. After around 300 EFC there are too few data samples available which explains the arbitrary mean SOH curve, i.e., predictions after that point in time should not be considered as valid.

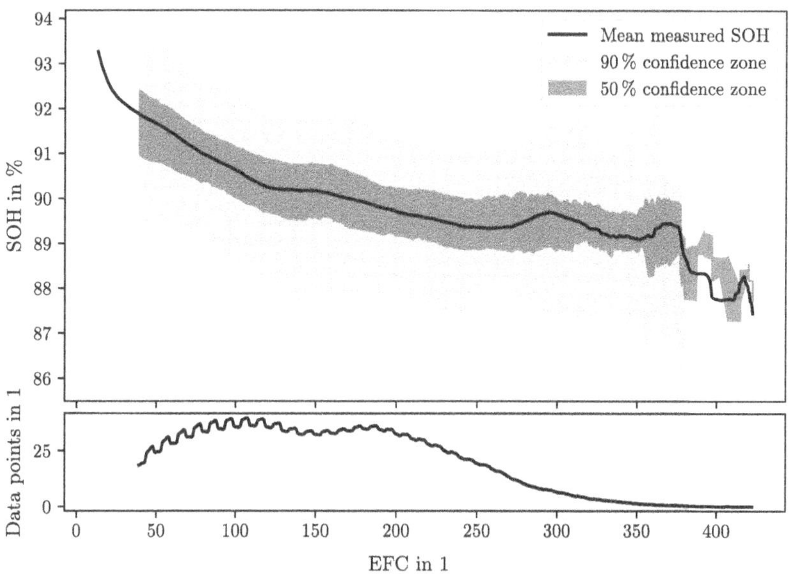

Figure 8.9 SOH forecasting error of the whole fleet with model H2 ($w_W = 25$ and $w_S = 10$). The characteristics of the mean measure SOH have been discussed at Figure 5.2 already

8.3.2 Accessible Features

For an initial assessment of the accessible features, Figure 8.10 depicts a boxplot diagram of the variance of the accessible features after min-max-normalization. The feature $\Delta k = w_W$ is always the same for single, i.e., non-grouped w_W, because the sliding window procedure is oriented at EFC and not, e.g., at the total operational time, e.g., with $w_\mathrm{w} = 30$ d. The features $I_\mathrm{ch,max}$ and SOE_max have a very low variance because their maximum value is always reached at some point during battery operation, either at DC-charging or after a complete charging session to $SOE = 100\,\%$ respectively. These features are expected to offer only little potential for differentiation of the samples during model training.

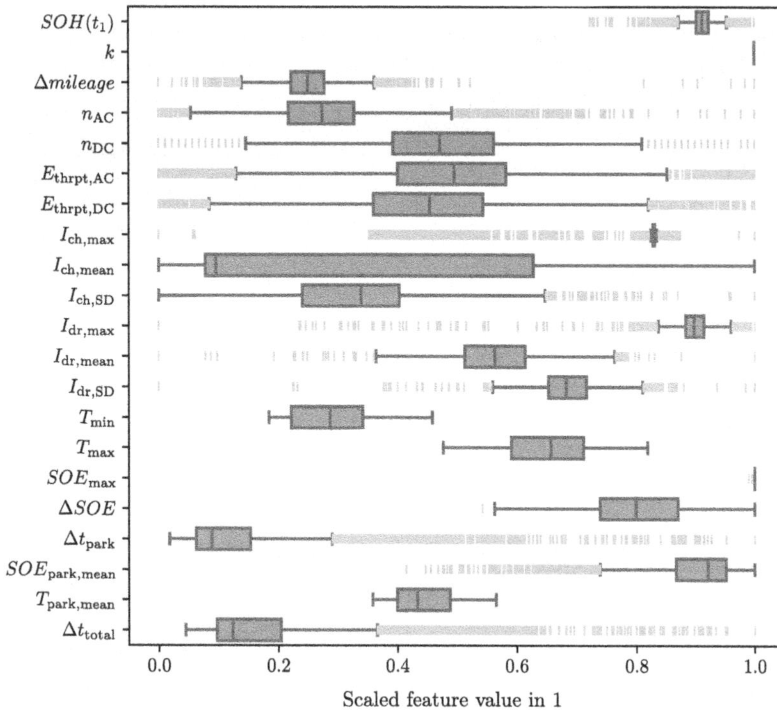

Figure 8.10 Variance of the accessible input features after min-max-normalization for $w_\mathrm{w} = 25$ and $w_\mathrm{s} = 10$

Experiment 1: Window shift (w_w)

Figure 8.11 depicts the model performance only with $SOH(t_1)$ as accessible feature for combinations of w_w and w_s. The model performance is quite similar for different w_s. Based on the evaluation criteria for the following experiments these four combinations of w_w and w_s are chosen: $(w_\mathrm{w}, w_\mathrm{s}) \in \{(10, 5), (25, 10), (50, 10), (100, 10)\}$.

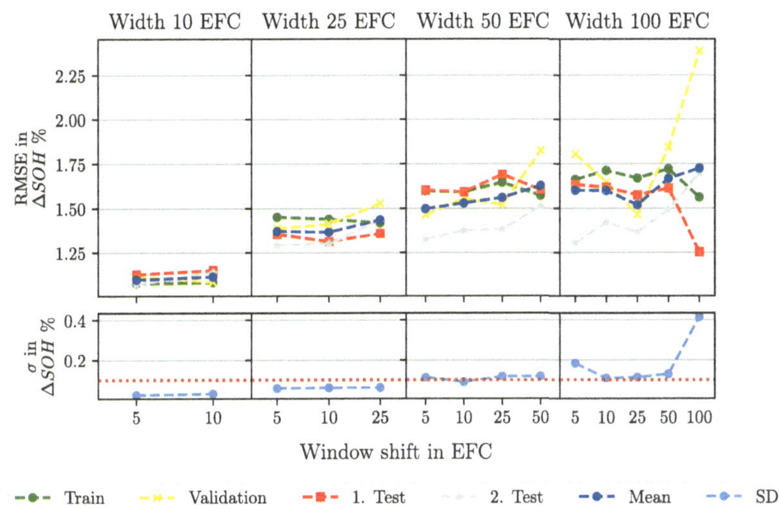

Figure 8.11 Model performance with different window shift w_s and only $SOH(t_1)$ as accessible feature

Experiment 2a: Cyclic Aging

The results of experiment 2a and 2b are compared to the base line of experiment 1 only with $SOH(t_1)$ as accessible feature. Figure 8.12 shows that the model performance with additional accessible features improves slightly. As for the histogram-based features SD increases for higher w_w, like $w_w = 100$ cycles. Feature sets 2a.4 and 2a.5 are performing best concerning the evaluation criteria ($E_{thrpt,AC}$, $E_{thrpt,DC}$, and $I_{ch,dr}$ respectively). Furthermore, they show a minor improvement of the mean RMSE compared to the baseline feature set 1 by 2.5 % and 1.6 % respectively.

Experiment 2b: Calendar Aging

As depicted in Figure 8.13, only the models trained on feature set 2b.4 fulfill the 1st evaluation criterion for all w_w. Also, for feature set 2b.4 (Δt_{total}) over all w_w the mean RMSE compared to the baseline feature set 1 decreases the most by 2.2 %. For feature sets 2b.1, 2b.2, and 2b.3, the decrease is only 1.5%, 0.5%, and 0.8% respectively.

Experiment 3: Cyclic & Calendar Aging

As depicted in Figure 8.14, when using all features from experiment 2a and 2b as feature set 3.1, the threshold of the 1st evaluation criterion exceeds for $w_w = 50$

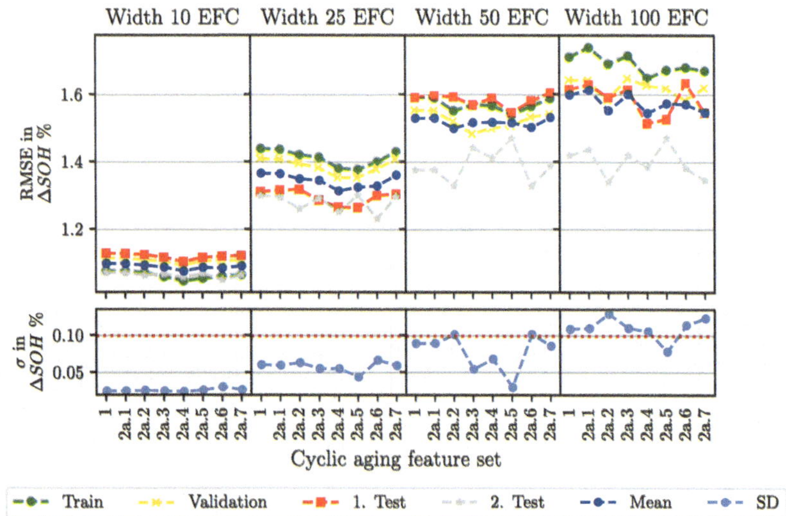

Figure 8.12 Model performance with accessible features describing cyclic aging

cycles. Using only the selected features ($E_{\text{thrpt,AC}}$, $E_{\text{thrpt,DC}}$, $I_{\text{ch,dr}}$, and Δt_{total}) from experiment 2a and 2b, together with the $SOH(t_1)$ in feature set 3.2 results in lower SD and improves the mean RMSE over the baseline feature set 1 over all window widths by 3.8 %.

Comparison of Histogram-based and accessible Features

Feature set 3.2 has only ten input features, which is significantly less than the 369 features of the histogram-based approach. Despite, the number of model parameters is roughly in the same dimension of 10^5, i.e., the model complexity is not always reduced. On average the mean RMSE is 2 % better for the histogram-based features compared to the accessible features over all w_{w} as listed in Table F.3 (accessible in the Electronic Supplementary Material). For a high w_{w} of 100 EFCs, the performance is similar because the operational load encoded in the histograms is smoothed strongly given the large aggregated time window. I.e., the operational load in the windows becomes similar. For smaller w_{w} of 10 and 25 EFCs, the histogram-based features are better than the accessible features (6.1 % and 5.0 % respectively). Thus, concluding that histogram-based features are preferable as they offer more detailed encoding and aggregation of battery operational load.

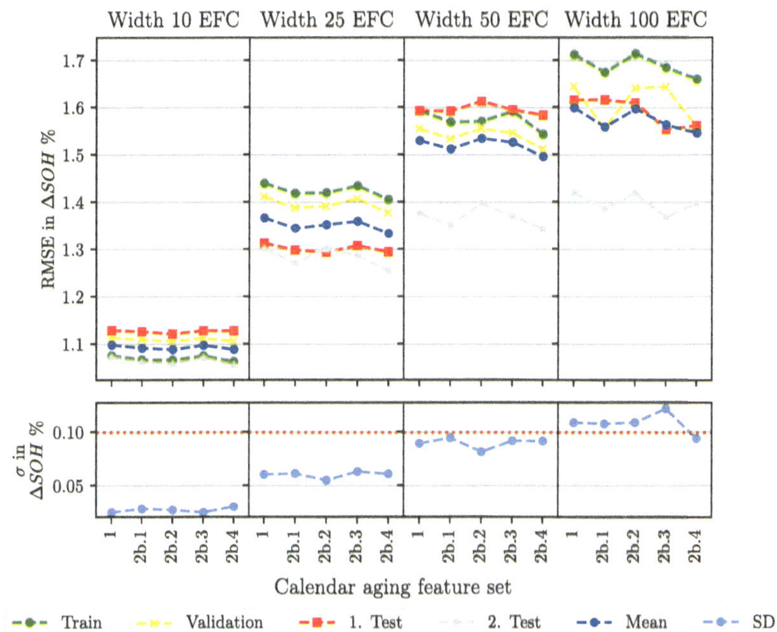

Figure 8.13 Model performance with accessible features describing calendar aging.

8.4 Use Cases

As implied in Chapter 1, when applying an SOH forecasting model, two variants to derive the input data exist: First, the model can be used for forecasting based on historical data. This would imply a forecast under the assumption of the same operational load as in the past. Second, the input features of historical data can be adapted, e.g., by fleet managers, under the assumption of certain battery operational scenarios as motivated in Section 3.3.

To exemplarily demonstrate how the SOH forecasting model could be applied, the second variant is applied in this section. Two use cases are applied to the model with histogram-based features (H2). For each use case the battery operational test data of the fleet is modified according to a scenario as introduced in Section 3.3. Therefore, the input data was fed to the model iteratively: The $SOH(t_2)$ of the next time window was calculated as $SOH(t_1)+\Delta SOH$ as suggested in Equations (3.1.2) to (3.1.4) of the previous time window with ΔSOH as the model output, similarly

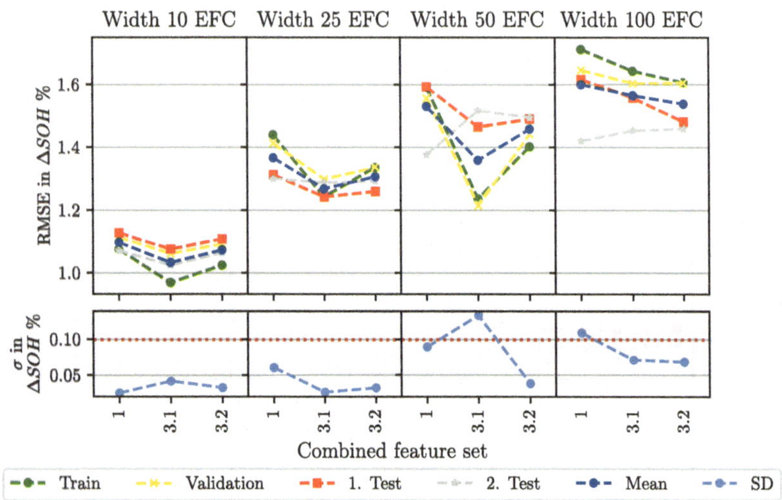

Figure 8.14 Model performance with accessible features describing cyclic and calendar aging.

to the suggestion of [403]. A drawback of this procedure is that the prediction error accumulates and the prediction performance decreases by time.[2] However, it allows to show the long term effect of the modified scenario data. All results in this section need to be interpreted carefully because the interfered results shown in the plots are only expressing what the ML model has learned, not necessarily how the aging really would have been.

Charging Management: Limitation of $SOE_{\text{park,max}}$

It is assumed that an intelligent charging management was employed to limit the maximum SOE at the end of a charging session. This reduced SOE during the following parking may lead to decelerated calendar aging. The mean SOE during parking for the baseline is at 90 % so that a limitation of $SOE_{\text{park,max}}$ to 90 % would only change the time series data a little. Furthermore, for the coarse signal interval width of 20 % these values would all remain in the same bin. Thus, it is not shown in Figure 8.15. Compared to the unmodified baseline, a $SOE_{\text{park,max}}$ limit

[2] Alternatively, a single samples could be split into several samples by linear interpolation of each histogram value x_i by k forecast sampling points: $x_{i,\text{subsampled}} = x_i \cdot n/k$ with $n \in \{1, 2, 3 \ldots, k\}$. For each k a separate forecast is run. However, this idea was disregarded because the ML model has not learned to forecast windows with the width of w_{w}/k.

of 80 % does not change the fleet's aging much. However, for a $SOE_{\text{park,max}}$ limit of 60 % the aging is significantly reduced. E.g., around 150 EFC the SOH is 0.8 % SOH smaller compared to the baseline. From there until 300 EFC the difference reduces, potentially because parking becomes less dominant after the breaks caused by COVID-19. The influence of limiting $SOE_{\text{park,max}}$ learned by the model and observed in this scenario corresponds to the results of Keil et al. [456] who find in laboratory cell experiments that calendar aging shows plateaus with SOC intervals of 20 % to 30 % with similar degradation, i.e., a non-linear dependency of SOC and calendar aging.

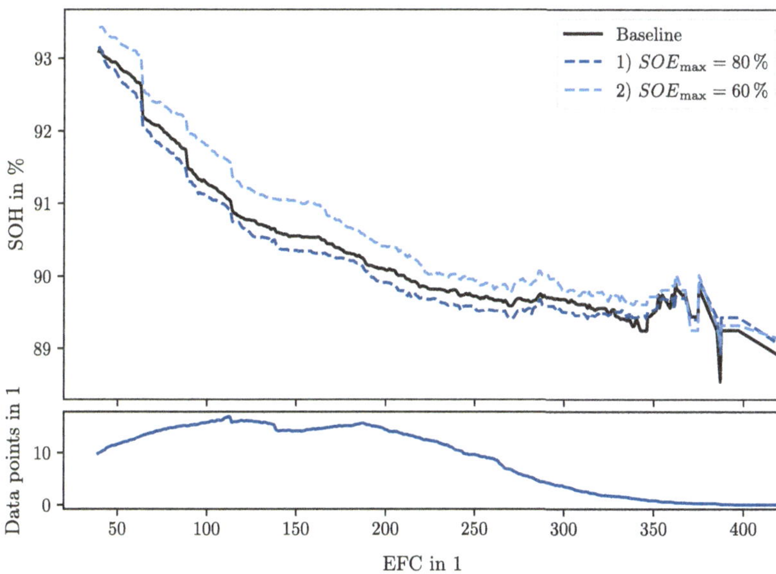

Figure 8.15 Fleet SOH forecasting with scenario of limited $SOE_{\text{park,max}}$ and model H2.

Charging Management: Only AC-charging

In laboratory cell experiments it was observed that high C-rates accelerate battery aging [57; 163]. In vehicle operation, high C-rates may occur during DC-charging, especially during HPC. Figure 8.16 shows the fleet aging trajectory of the unmodified baseline scenario with a mixture of AC- and DC-charging. When assuming AC-charging by limiting the C-rate, the SOH increases by about 0.1 % SOH which was considered negligibly small. However, the DC-charging C-rate was only around

1.25 C during DC-charging, so relatively low compared to battery cell experiments with up to 8 C [69; 70]. Also, it has to be noted that for switching from AC- to DC-charging the duration of the charging session, temperature, and SOE would change. This was not done in this scenario so that only the influence of changed current distribution is considered, i.e., ceteris paribus. Further influences, e.g., of charging at lower and high temperatures were not considered in this use case.

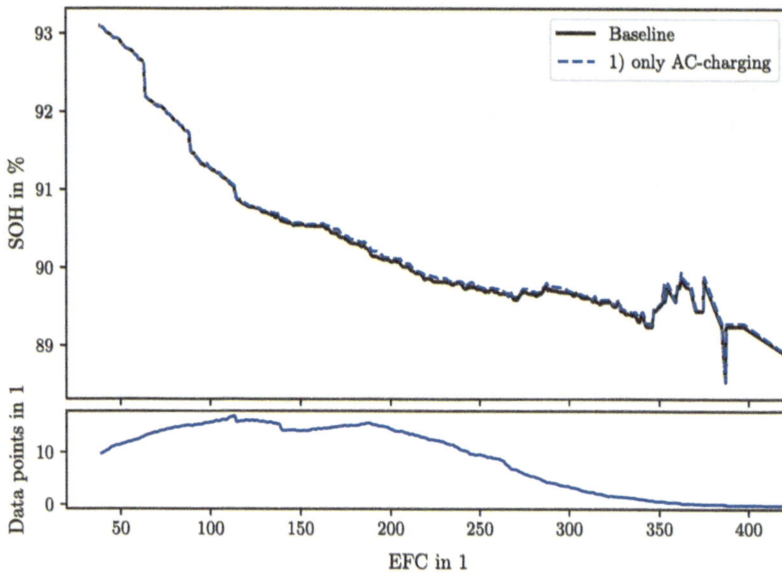

Figure 8.16 Fleet SOH forecasting with scenario of only AC-charging and model H2.

8.5 Summary and Contribution to the Research Questions

In Chapter 6 and by [1; 392] it has been shown that SOH forecasting works with laboratory battery cell data. This chapter successfully applied SOH forecasting based on histogram features on battery system data from real-world BEV fleet operation of a ride-pooling MOD provider (RQ.3). The model was able to learn the dependence of the SOH from the battery load, i.e., BEV usage. Compared to the laboratory battery cell data, generalization of models trained on vehicle system data did not improve when learning multiple forecast horizons, i.e., w_W. Also, overfitting on vehicles freshly joint to the fleet was controllable by the amount of samples generated, i.e., w_S. Two use cases for different charging strategies exemplary show how the SOH forecasting model may be applied. Switching from accessible features to the histogram-based features showed an improvement in model performance of up to 6.1 %.

9 Concept for a Technical Implementation

This chapter discusses RQ.4: „How can SOH forecasting of LIBs in automotive applications technically be realized?“ To answer this question also necessary infrastructure and systems are discussed especially from the perspective of the BEV manufacturer. This chapter is based on previous publications [6; 7].[1]

In the following, this definition of the term „BEV fleet“[2] is valid: A BEV fleet is a group of BEVs, that is homogeneous with respect to their function, clustered by certain criteria, which may be technical, operational or contextual. Possible criteria are the ownership, the geophysical location or region of operation, the type of user or operator, the model type or generation of the fleet members, as well as the aging or degradation state of the fleet members.

[1] For [6] © 2021 IEEE.

[2] In the scope of this thesis, the focus lies on BEVs, but other machines can also constitute a fleet as motivated and defined more broadly and extensively in Appendix G.1 (accessible in the Electronic Supplementary Material). There, also existing definitions of the term „fleet“ are analyzed.

Supplementary Information The online version contains supplementary material available at https://doi.org/10.1007/978-3-658-43188-4_9.

F. von Bülow, *A Data-Driven Fleet Service: State of Health Forecasting of Lithium-Ion Batteries*, AutoUni – Schriftenreihe 170,
https://doi.org/10.1007/978-3-658-43188-4_9

9.1 Battery Electric Vehicle Fleets

One major cost driver of the development process of batteries is testing [371; 457]. To guarantee the longevity of battery cells, endurance testing consumes time and resources, but is mostly restricted to limited testing facilities. Once series production commenced, frequent workshop measurements of the battery's SOH would be necessary to monitor battery aging due to vehicle usage. Unfortunately, this procedure is unfeasible for workshop facilities and customers. In order to make battery testing scalable, to lower the cost, and to reduce battery testing efficiently, simulations with digital battery twins show great potential [52; 351; 386; 458]. Such digital twins can be suitable for SOH estimation, like the model of Heinrich et al. [52] who perform standard laboratory battery tests virtually with almost experimental accuracy using only real in-vehicle driving data for training.

Such models can be trained inside the vehicle during operation. Hence, the whole battery functionality can be compressed to a minimum amount of model parameters. This compression enables an efficient and scalable data transmission between single vehicles and the fleet management.

The fleet operator can thereby analyze each member of the fleet individually. Comprising the knowledge of all operated fleets, optimization strategies, predictive maintenance, and adjustments for the battery operational parameters can be derived. This procedure benefits the customer's performance and can be further used to improve development of products and services. Understanding the real utilization of vehicle components by a variety of different customers, the development of new model generations can be optimized in terms of requirement specifications, prevention of over-engineering, and new exploitation concepts.

9.2 Stakeholder Roles of Fleets

Until now, few authors have analyzed fleets' stakeholder roles involved in operation and manufacturing of fleets. However, this is essential for a holistic fleet management approach because fleet management serves a purpose of interaction and support of the fleets' stakeholders.

Michau et al. [459] consider fleets from the perspective of the manufacturer and operator. However, the operator role needs to be refined further. It may refer to one or several persons operating the machine, for instance the driver of a vehicle or an operator team of a chemical plant, but could also refer to the institutional fleet operator like a company, which usually will also own the fleet. Hence, when operating a fleet, different stakeholder roles can be involved: These are the owner, the fleet manager, the operators, like machine or plant operators, workers or drivers, and maintenance crews. For fleets these roles are usually neither filled by the same person or team nor are these necessarily part of the same organization.

In the following, possible tasks of each role are described. The owner, usually an institution, is responsible on the strategical level. This includes deciding on the fleet's task, the fleet member selection, and the fleet size. Regrading investment decisions the holistic methodology of the total cost of ownership (TCO) is usually used [460]. The TCO considers all the costs that occur during the ownership of a capital good, like a machine or vehicle. It includes not only the cost of acquisition, but also costs during lifetime and usage, e.g., for energy, repair, and maintenance.

The fleet manager takes responsibility for the operational daily business. They manage accidents, service stops, schedule shift plans, and monitor the overall fleet performance and usage. In the case of BEV fleets, the fleet manager also supervises the charging management, e.g., by selected preferred charging profiles.

The operator is working at, with or on the fleet members on the operational level and is responsible for a fleet member's usage. Examples are a vehicle driver and a team operating a steam cracker in the control room supported by a process control system (PCS).

These roles might not be separated or existing for all types of fleets. For example, on the one hand, a fleet of windmills is operated fully automatically and won't need human operators, but will still be monitored remotely by a a supervisory control and data acquisition system (SCADA-System). On the other hand, a fleet constituted of all vehicles of a certain vehicle model type registered in Germany currently neither has an operational fleet manager nor a joint owner. As each driver or a group of drivers owns one vehicle, no fleet manager as intermediate exists. Contrarily to that, in the case of a non-autonomous ride hailing vehicle fleet, all roles of owner, fleet manager, and driver exist non-separated.

9.3 State of the Art: Fleet Management

No fleet management approach integrates the manufacturer's perspective to support the development process at the moment. Currently, there exists a wide range of fleet management systems supporting the aforementioned roles of operators and fleet managers. These systems are often offered by manufacturers or third-party companies for different types of machines like floor care equipment [461], intralogistics solutions [462], and especially for vehicle fleets of company cars, leasing cars, and logistics fleets [463-468]. Simple telematics systems provide fleet monitoring for fleet managers, e.g., regarding fuel and energy consumption, fleet position, and mileage [469–471]. These monitoring systems only transfer data, aggregate it, and display it to the fleet manager [470; 472; 473].

More advanced fleet management systems support fleet managers and operators by integrating operationally useful components like dynamic vehicle routing, maintenance management, and cost management [474–480]. Especially for BEV fleets, charging management approaches have been proposed [481; 482]. Not only for vehicle fleets, but also for other fleets predictive maintenance approaches have been considered [83; 483–485]. However, these approaches focus only on a single fleet, not allowing the transfer between several different fleets.

For the BEV fleets relevant to SOH forecasting, manufacturers only have limited data of new vehicles, e.g., from test vehicles and endurance testing (L4). Albeit, especially the latter is immensely expensive. Additionally, both only supply a limited data base regarding quantity and variety. They are usually executed by a group of test drivers and do not reflect operational usage of end customers (Key criterion 1c). Thus, a holistic fleet management approach from the manufacturer's perspective will support and reduce costs of the development process.

9.4 Fleet Management Concept for Manufacturers

Due to the limitations of SotA fleet management system for manufacturers, this section proposes a fleet management approach for manufacturers. First, the general layer structure is introduced following a detailed description of the layers and the information flow in between the layers in the next three Sections 9.4.1 to 9.4.3. In Section 9.4.4, possibilities for the data storage and ML model training are discussed.

The proposed concept has a structure of three layers which is depicted in Figure 9.1: The machine, fleet operator, and manufacturer layer. Like the automation pyramid [486–488], the layer structure reduces complexity. The automation pyramid consists of four layers: Enterprise layer, plant layer, process layer, and field layer. Its objective is the production control of plants or machines [486–488]. Even though the proposed concept has a different objective, it is related to the automation pyramid. However, this concept puts another layer on top of the automation pyramid: The manufacturer layer. The automation pyramid's enterprise layer is comparable to the fleet operator layer as it refers to the (fleet) operating enterprise. Additionally, plant, process, and field layer are merged into the machine layer because their exact structure is less relevant for the fleet management.

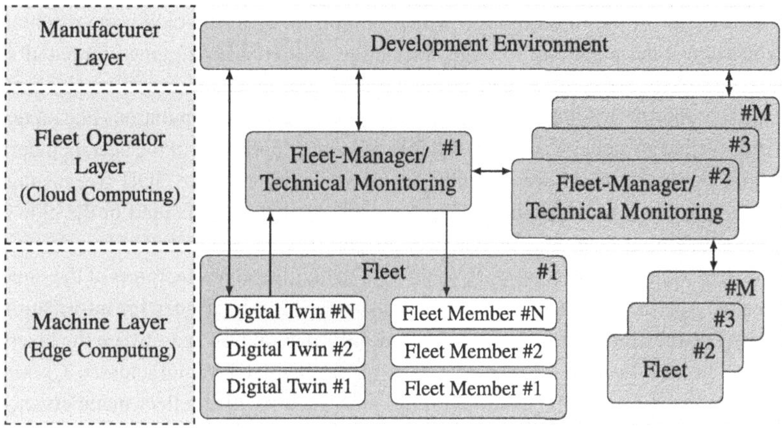

Figure 9.1 Fleet management concept for manufacturers. [6; 7] ©2021 IEEE

On the bottom machine layer of this concept, each fleet is constituted by grouping machines regarding certain criteria following the chosen fleet definition. Thus, single machines are viewed as fleet members. Each machine is accompanied by an individual digital twin. The digital twin mirrors the life of its corresponding physical twin that can be used for various purposes [489]. The objective of this layer is to support the single machine's operator, operating teams, and maintenance workers, so it is machine-focused.

On the middle fleet operator layer, each fleet is supervised by the roles of technical monitoring and fleet manager. The technical monitoring focuses on the fleet member's health and load from a technical point of view. The fleet manager takes responsibility for the operational daily business like described in Section 9.2. Depending on the technical expertise and need of the fleet manager, the roles of technical monitoring and fleet manager can be executed jointly or separately, either by a person or a team. At all events, for larger fleets that are not operated by a single person or institution technical monitoring and fleet manager can be roles allocated at the manufacturer, e.g., in the after sales or service department. The objective of this layer is to support fleet-related activities of the fleet operator, so it is focused on fleet operators.

On the top manufacturer layer, all technical monitoring and fleet managers from the fleet operator layer are connected to the manufacturer's development environment. The objective of this layer is to support the manufacturer of the fleet members independently of the composition of the fleets, so it is manufacturer-focused.

In contrast to the middle layer, which is centralized in the cloud, the machine layer follows the paradigm of edge computing as it enables moving computation and data storage away from a central cloud [490]. This division is motivated by cost, security, connectivity, and bandwidth limitations. Data and computations like model execution that are necessary for the reliable and secure operation of the fleet members stay on the machine. However, additional digital services, like SOH forecasting, provided to the fleet's stakeholders and model training are executed in the cloud. The manufacturer layer may be realized independently as the manufacturer desires.

When extending this concept from a single to multiple manufacturers of the same machine type, interchangeable layers with standardized interfaces for information flow in between the layer become an option. Information flow in between the layers will be limited by data access directives of the involved manufacturers. Consequently, fleet data is only shared with the manufacturer of the fleet members, not with other manufacturers. Unless, other learning paradigms like federated learning are implemented. Also, of all layers only the fleet operator layer can realistically be exchanged to limit external access of third-party manufacturers to critical functions.

In the following Sections 9.4.1 to 9.4.3, the three layers are described in more details at the example of BEV fleets.

9.4.1 Machine Layer

On the machine layer, each fleet member's digital twin is composed of individual information regarding the components of interest. Depending on the complexity of the aging causes of the component either component statistics are recorded or intelligent on-board models are used. Component statistics encode how the components have been stressed during usage. For BEVs this could include simple metrics as mileage or the number of charging cycles. Component statistics may suffice for operating materials, tire wear, and components of the drive train.

More complex aging mechanisms can be captured by intelligent on-board models. For BEV fleets such complex aging mechanisms exist in the battery as introduced in Sections 2.1.4 [458]. Intelligent on-board models may be physical, hybrid or ML models. However, physical models require complex model development for each new model. Thus, hybrid and ML models are beneficial. Nonetheless, they are dependent on data gathering from fleet usage. This data may also serve for validation of physical models.

The definition and adjustment of the component statistics and intelligent on-board models is done by the manufacturer from the highest layer. This enables interoperability and comparability of the digital vehicle twins. Furthermore, the task of designing digital twins would be too complex and out of scope for fleet operators.

9.4.2 Fleet Operator Layer

On the middle fleet operator layer, the roles of technical monitoring and fleet managers are supported by the technologies of intelligent data aggregation, fleet analysis toolbox, predictive maintenance, and data driven optimization methods. Access to these functions depends on the technical expertise and need of the fleet manager.

Like in SotA fleet management systems, the fleet manager is supported with a fleet analysis toolbox that provides operationally relevant information like current states of the fleet's vehicles and their components. In the case of BEVs this would be, e.g., the SOC and SOH. For fleet analysis, the manufacturer can provide services to the fleet operator over the entire product life cycle like RV estimation and SOH forecasting. Also, interfaces to third-party providers are possible on the fleet operator layer.

Furthermore, the data from the component statistics and the intelligent on-board models gets aggregated component-wise, for each fleet member, and the entire fleet in the fleet data storage for displaying in the fleet analysis toolbox. This data is relevant for the fleet manager, but also for the technical monitoring of the fleet. In this way relatively fast aging components within the fleet can be identified, analyzed, and potentially maintained.

The fleet manager has the capability to customize the control and operational parameters of the fleet members related to the operational strategy. Hence, the fleet manager can implement measures to extend the lifetime, minimize operational costs, and maximize the fleets performance or utilization. Before updating customized control and operational parameters to the fleet, the technical monitoring and the fleet manager are supported by data-driven optimization models. As motivated in Section 3.3, these models enable scenario-based simulations to identify the desired impact of the chosen control and operational parameters beforehand. For BEV fleets customized control and operational parameters could specify the configuration of the low power mode at low SOCs.

On the middle fleet operator layer, information exchange in between the different technical monitoring and fleet managers across fleets is possible, e.g., implemented as model transfer as suggested in Section 3.4 and extended to BEV's apart from the battery model transfer in Appendix G.2.

9.4.3 Manufacturer Layer

On the top manufacturer layer, fleet data stored for each fleet on the fleet operator layer is aggregated fleet-comprehensively in a global data storage.

This global data storage may serve as early input to the development process of new products and services because it contains information of the real usage behavior of several fleets. Currently, such information may only be acquired qualitatively by customer survey indicating customers' needs, but not the customers' real usage of the machines. Especially, data of overloading and common load scenarios are interesting for the development to specify new machine models. Consequently, the product requirements document (PRD) of future components will be based on actual customer demand. This also applies to the interaction of the components as complete product. Based on the cross-fleet historic data storage, the manufacturer can develop customer-demand orientated services for hardware and software components.

Additionally, development engineers can interact with the fleets directly to roll out fleet-wide OTA software updates. An example are model adjustments of the underlying structures of the fleet members' digital twins. This improves the information basis for fleet managers provided by the digital twins. Development engineers can also prototype functions on a sub fleet to get fast customer feedback [491]. Additionally, shadow testing [492] which is known in the software domain can be applied on the fleet when connected to the development department. This is especially relevant for software functions and digital services provided to customers, drivers, fleet managers, and fleet operators.

2nd life applications of fleet members or their components usually need to be scheduled to meet supply and demand timely. This applies for example to BEV batteries transferred to a 2nd life applications for stationary energy storage. Also, components with high RV for recycling may be observed closely at their EOL to improve scheduling the recycling plant. Both use cases are interesting for the business model of batteries as a service (BaaS) [493].

9.4.4 Data Storage and Machine Learning Model Training

The question of the location of data storage and ML model training depends on multiple factors like the origin and quantity of the training data and the computational power needed for training. For example, for minimizing the data transfer the intelligent on-board models could be trained on the edge device where the data is located. Contrarily, fleet-comprehensive models that require training data of all fleet members of a fleet can be trained either centralized in the cloud or federated, as suggested in Section 10.4. The latter is advantageous, when the data interface between the fleet members and the cloud causes privacy issues or suffers from unreliable and slow network connections. Privacy issues and General Data Protection Regulation (GDPR) concerns may occur when data regarding the detailed usage of the fleet members is transmitted. For vehicle fleets, this includes information like the location, driven routes, and periods of use. To overcome these issues federated learning may be applied in a fleet context.

Federated learning is an ML paradigm on training a central model while the training data stays distributed at many clients, i.e., fleet members in this context. Based on a central model provided by a central back-end, each fleet member independently computes a parameter update to the current model using its local data. The fleet members only communicate the parameter updates to the central sever, where all updates are aggregated to a new global model [494–497]. In contrast to federated learning, distributed learning assumes the availability of all training samples in a centralized location. From there data can be shuffled and distributed over computation nodes [496]. In this case, privacy issues could be tackled by anonymizing the transmitted data.

From a technical point of view two variants of data pipelines for SOH forecasting split into on-board and off-board aggregation exist:

1. Raw time series data could be transferred from the fleet vehicles to a central back-end could and there aggregated features like histograms could be derived off-board (Manufacturer layer).
2. The aggregated features could be derived directly on-board (Fleet Layer) and only then be transferred from the fleet vehicles to a central back-end.

Variant 2 has the advantage that cost for data transfer is smaller because the amount of data transferred is smaller than for variant 1, especially for high sampling rates and multiple time series signals. However, variant 1 is advantageous over variant 2 because stressor tables with different bin sizes can be derived in the central back-end. Also, a more detail analysis of the time series data, e.g., regarding certain operational situations is possible. For variant 2, transfer of the aggregated data may be executed after one or several EFC, after one or several days.

The results of Chapters 6 and 8 show that w_{w} and w_{s} have a major influence on the goodness of fit of the ML model. Thus, it would be advantageous to maintain a flexible selection of w_{w} and w_{s} even in the case of variant 2. Therefore, a minimum w_{w} like 1 EFC can be set after which the aggregated features, SOH (t_1), and SOH (t_2) of that cycle or time period are send to the central back-end. This enables later aggregation to larger w_{w} with variable w_{s}, both as multiples of the minimum w_{w}, in the central back-end with an extensive examination.

Limitations & Outlook 10

The answers to the RQs developed and presented in Chapters 6 to 8 are critically reflected in this chapter. Based on their limitations possibilities for future research are presented. Overall, there is no single direction to which future research should commit. Given the novelty of SOH forecasting with real-world automotive operation data, there are rather plenty of possibilities from which to chose for researchers and practitioners depending on the available data and the application scenario of future SOH methods.

First, Section 10.1 focuses on limitations and further possibilities for the presented method given the data sets used in this thesis. When applying the public laboratory data presented in Section 5.1, these proposals may be implemented by any researcher. Going beyond the given data sets, Section 10.2 presents possibilities with further data whose availability might be limited for researchers. Some of these data sources may only become available in the future once certain battery technologies like solid-state-batteries (SSBs) are more mature. As elaborated in Section 10.3, the same applies to future research once more detailed data of battery systems becomes available. Beyond transfer learning which has been applied to cell data from laboratory, potential benefits of further learning paradigms known in the field of ML for SOH forecasting are discussed in Section 10.4.

F. von Bülow, *A Data-Driven Fleet Service: State of Health Forecasting of Lithium-Ion Batteries*, AutoUni – Schriftenreihe 170,
https://doi.org/10.1007/978-3-658-43188-4_10

10.1 Method

Path Dependence

The current battery state is only described by $SOH(t_1)$. This could be problematic in the case of capacity regeneration, i.e., an increase of the SOH during rest periods due to reversible capacity loss [164], because one battery may reach one SOH several times during its lifetime. Also, Baure et al. [498] point out, that battery cell aging is more complex than what capacity and resistance reflect. Thus, for future work, the current state of a battery could be expressed by more variables, i.e., as a vector as suggested in Section 2.1.3. Apart from SOH_C, for example E_{thrpt}, the BEV's mileage, and the share of rest time in the recent past could enable a further differentiation of the current state.

Delayed degradation effects caused by past battery operation with $t \leq t_1$ are neglected, i.e., path independence before the forecasting time window $[t_1, t_2]$ is assumed. This is coherent with key criterion 1a to focus on the future battery operation which is not yet determined. To consider delayed degradation effects, the past battery operation or the past SOH trajectory could be considered as model input. However, this has the risk that past and future battery operation get mixed up.

Further, path independence is assumed in the forecasting time window $[t_1, t_2]$ because no sequential information about the battery operation in the forecasting time window is encoded in the histograms or accessible features [398; 498]. However, this is a limitation because the order of operational states encoded in the histogram matters [167; 168].

To consider path dependency within the forecasting time window, this overall forecasting time window can be split into sub-windows which could span for instance only $w_{w,s} = 1$ cycles while the forecasting time window is bigger with for instance $w_w = 25$ cycles. For each sub-window n_1 features can be derived just as they are in the proposed method in this thesis. Due to the sliding sub-window, a sequence of aggregated sub-windows is retained which aggregates the original time series. This means, the input dimension of n_1 features from the proposed method is extended to $n_1 \times n_2$ with n_1 features for each of the n_2 sub-windows. This adaption of the preprocessing does not require adaption of the training sample generation with w_w and w_s described in Section 6.1.1. Also, the output value ΔSOH would remain the same. The features of the sub-windows can be statistical parameters like arithmetical mean, SD, 25th, and 75th percentile of a time window [334]. Alternatively, like proposed in this thesis, histograms can be used. The sequential histograms of dimension $n_1 \times n_2$ can be input of an LSTM to consider the timely sequence of the operational states at least on the level of the sub-windows, while still avoiding vanishing gradients of LSTMs. When sub-window sampling is too fine for RNNs similar to NLP problems for long sentences [499], attention-based ANNs may be able to learn relative importance depending on the sequence and values of the histograms.

Variable Binning
In the current method, the interval limits and range of the histogram bin's are fixed and require manual tuning depending on the operational domain of the batteries. In future research, variable binning of the time series data could be achieved by converting the binning to a learning problem itself. Variable bins could sample important operational state regions of the batteries more finely and aggregate uncritical regions more. Also, dynamic bins, i.e., variable bins over time, could sample the recent operational data more finely and aggregate past data more.

Time Series Aggregation
This thesis applied 1D, 2D, and 3D histograms as well as certain more accessible features as model inputs. However, there are further possibilities to aggregate time series of I, T, and SOC: Similar to [500], operational patterns of the BEV and battery usage could be identified unsupervised and their frequency of occurrence could be used as model input, e.g., 10 launch controls. When disregarding the need to adapt the battery operational data, or histograms to run scenarios in production, BiSCo could provide efficient and lossless compression [78].

Sequence Output
A single forecasted SOH may be helpful in comparison to other SOH values, but often a complete forecasted SOH trajectory provides more contextual information of the forecast. Also, an SOH trajectory is more easily verifiable for consistency. For the use cases presented in Section 8.4, a forecasted SOH trajectory, is obtained by adapting histograms of several shifted windows and running multiple model interferences with those adapted samples. In future research, an ML model can learn an SOH trajectory directly. This means that the output dimension is extended from 1 to n_2. For example, k sampling points of the SOH within the time window could form an output sequence. Alternatively, when using the sub-window approach proposed earlier in this section to consider path dependence within the forecasting time window, the SOH after each of the n_2 sub-windows could be output of an RNN transitioning to a sequence to sequence (seq2seq) problem.

Hybrid Method
This thesis applied a purely data-driven model which is limited by its training data. However, it may be beneficial to set up a hybrid method which combines knowledge gained from laboratory experiments, e.g., the Arrhenius law for calendar aging [369]

and battery data from the operational domain [72, pp. 363–364]. For example, in physics-guided neural networks (PGNNs) a loss term for physical consistency is added to the loss function [501]. The loss term is based on equations describing the physical relationship between the output variable y, in this thesis the ΔSOH, and other physical variables. A major problem will be to transform physical laws to consider distributional data, for instance, not only one SOC_{storage}, but several or a distribution of SOC_{storage}. Hybrid methods would have more inherent white box character, i.e., being gray box models, supporting explainability and interpretability of the forecasts as mentioned below in Section 10.4. Possibly, less data is required for their training as for purely data-driven models.

Application in Optimization

In the presented use cases in Section 8.4, the samples were adapted manually according to the defined scenarios. Going a step further, this adaption could be done by an optimization algorithm if the SOH forecasting method is embedded into an optimization problem $\min_x f(x)$ with the inputs of the ML model as variables x similar to [502]. This would enable consideration of the learned information on the battery aging resulting from a chosen battery operation during solving a continuous optimization problem. An optimization objective like the TCO of a fleet could be minimized given operational constraints like the fleets availability and charging possibilities expressed as inequality constraints $g_i(x) \leq 0$ with $i = 1, \ldots, m$ and equality constrains $g_j(x) = 0$ with $j = 1, \ldots, p$ [503]. Other minimization objectives could be the fleet's total SOH, but also the maximization of homogeneity or heterogeneity of the fleet's SOH measured by the SD of the SOH of the fleet's BEVs [12].

10.2 Future Data

Optimal Data Set

The data sets presented in Chapter 5 limited the presented findings. As a major model input and currently single output value the SOH and its measurement accuracy as well as resolution are a major limitation. Also, the sampling time may pose a limitation. This is the case for the BEV fleet data used in Chapter 8. Especially, the sampling time of the current signal in drive mode is critical. Additionally, too few measurements of the temperature during parking could limit the model's performance as calendar aging is connected to the temperature [160; 161].

Thus, in future work it seems advisable to get access to an ideal training data set for SOH forecasting as specified in the following. BEVs should be driven significantly so that the batteries have aged. A minimum of 10^5 km driven or 5 % of SOH degradation seems advisable. Otherwise the battery degradation is too small to be learned by the SOH forecasting model. The quantity of vehicles or batteries required depends on the variability of the operation and technical systems. More variability requires more sampling points, i.e., more vehicles or batteries.

As stated in Section 5.2, the histograms derived from the BEV fleet data are limited by their sampling rate which only approximates the probability distribution of the underlying higher-sampled time series [401, p. 5]. Thus, Table 10.1 specifies the desired sampling rate of an ideal data set. Voltage and current require the highest sampling rate because they are the most dynamic signals. For the temperature measurement during parking mode, it may be important to take regular measurements, e.g., every hour. This ensures capturing the dynamic of the environment temperature in the course of the day.

For validation of the SOH values, reference measurements at a service station may be beneficial.

Table 10.1 Desired signals for SOH forecasting model.

Signal	Ideal sampling time
V_{cell}, V_{pack}	100 ms
I_{cell}, I_{pack}	100 ms
T_{cell}, T_{pack}	On value change (1 s)
SOC	On value change (1 s)
Battery, vehicle mode	On value change (AC-/DC-charge, discharge, hold)
Capacity, SOH, E_{max}	Daily/weekly

Data on Mechanical Load

The data used in this thesis originates from cylindrical cells. Pouch cells are another common cell format. As shortly described in Appendix A.1.2, also the cell's format influences the degradation behavior. Especially, for pouch cells, aging is influenced by the clamping force which would require additional features for SOH forecasting like the stack pressure [173; 504].

In non-stationary applications, LIBs especially in BEVs like in Chapter 8 may also face mechanical stress through vibrations [344; 505–507]. This could be included in the model inputs through further features which could be derived from the accumulated longitudinal and lateral acceleration of the BEV's body that the battery also faced. Alternatively, additional sensors directly within the battery system could provide such information. These sensor signals from the CAN could be processed to histograms, possibly adding a further dimension to the existing histograms, or analogous to the accessible features as maximum stress values. Also, the frequency spectrum of the vibration may be used as a feature after extraction by FT or wavelets [443].

Data Augmentation

Data augmentation techniques improve the size and quality of training data for DL models. For example, basic image manipulations such as flipping, cropping, rotation, and translation have shown good results to add variance to the training data and help avoid overfitting [435].

Data augmentation for SOH forecasting training data set is considered to be difficult for several reasons: Multivariate time series signals of time spans of at least several days influence the SOH trajectory. Especially, in the case of NI-VC battery operation the functional connection of the SOH trajectory and the battery operation is too complex to be comprehensible for a human. Thus, even the influence of simple adjustments of the SOH trajectory or the battery operation is unknown. Such adjustments could include changing the order of discharge-charge cycles, limiting battery operational parameters, or extending the residence-time in certain operational states. Also, creating synthetic SOH trajectory by adding an offset, changing the slope, or adding elongation lacks the underlying change of the battery operation for the applicability to SOH forecasting. Thus, it is not applicable and beneficial for the presented SOH forecasting method.

Metadata and Text Data

In this thesis, each battery data set, i.e., each battery type and operation, was used to train a separate ML models for SOH forecasting. In another previous work [4] from the context of this thesis that is not further presented in this thesis, the laboratory cell data presented in Section 5.1 is jointly used to train a single ML model. However, the ML model needs to learn metadata of the different batteries like battery materials, voltage limits, and temperature ranges or operational climate zones implicitly from the operational data. When training a single SOH forecasting model for several battery types it might be beneficial to explicitly provide information about inherent differences of the battery types, e.g., as categorical model input.

Currently, the majority of SOH forecasting models including the method presented in this thesis only considers time series data or features derived from it. Text data like technical annotations from workshops are not considered. Such information could include information about certain parts of the battery like the battery junction box (BJB), BMS, or certain modules were maintained, replaced, or in the case of software updated at a given point in time. A first step into the usage of such textual maintenance data is the work of [508] who apply pretrained NLP models to process technical language.

Future Data Sources

In the future, an SOH forecasting model could also be build on information accessible in the European Union (EU) Battery Passport. E.g., it is planned to include "the number of charging and discharging cycles, and negative events, such as accidents" [509, Annex XIII, part B]. Even more important for SOH forecasting also included are "periodically recorded information on the operating environmental conditions, including temperature, and on the state of charge" [509, Annex XIII, part B]. Details are to be defined in delegated and implementing acts. The frequency of the periodical records may limit the data's informational value about the operational load.

Currently, new battery cell concepts such as SSB and lithium-sulfur battery (SIB) cells are in the focus of research. In the future, these, but also other concepts may become relevant for BEV series production [510].

In SSB the liquid electrolyte is replaced by a solid material. This suggests that aging of SSB may be triggered by similar operational stressors. Indeed, reducing the discharge rate decelerates aging [511, p. 10]. But different to a LIB cell with liquid electrolyte, increasing the operating temperature of a SSB cell, e.g., from 20 °C to 40 °C, decelerates aging [511, p. 12]. Overall, SSB can withstand higher temperatures than cells with liquid electrolyte [512]. Because of these fundamental similarities the application of the proposed method in this thesis to SSB seems feasible once sufficient battery operational data of SSB is available.

As BEVs also heavy duty (HD) battery electric trucks (BETs) are becoming commercially available [19, p. 35]. Here a major change will be the megawatt charging system (MCS) that shall charge the BET with higher powers than 350 kWh like for HPC of BEVs, e.g., up to 1 MWh and 4.5 MWh are discussed [513, p. 36; 514]).

Batteries build in trains, i.e., BEMUs, are currently rare, but offer advantages for examples on partially electrified railway networks. In such applications the cumulative stress for the batteries is estimated to be 20 times higher than in BEV's because of a higher mileage per day, less parking time, and a longer lifetime [515].

SOH forecasting of batteries in these two applications offers potential for optimization of the operational strategy because of these more demanding requirements. In both application the changed battery operation compared to BEV would require special consideration. The selection of the histogram's bounds should be adapted to accurately encode the higher C-rates. For BEMUs with less parking time, different normalization of the corresponding histograms seems advisable.

10.3 Battery Systems

When assuming data at system-level and cell-level of BEVs' battery systems, the interdependencies of current, temperature, and SOC on the cell's battery aging when switching from cell to pack may be considered in more detail as discussed in Section 2.1.5.

For example, the applied method in Chapter 8 only uses the one available temperature signal of the system. In future work, the input feature set could be extended for battery systems when several temperature signals are available, e.g., at module- or cell-level. Even further, not only the temperature, but also current, and SOC at cell-level may enable an SOH forecasting model to track the SOH of each cell, i.e., the SOH distribution of the battery system. Cell-to-cell voltage deviations were also observed to have strong impact on the SOH [516]. They reflect a need for cell-balancing as introduced in Section 2.1.5 and may also be considered as inputs. Such a model could especially consider the aging interdependencies of all cells as described in Section 2.1.5.2.

Currently, the intrinsic cell-to-cell variance due to production, i.e., the cells' initial SOH, are not considered by the presented method. However, this could be beneficial given the study of Baumhöfer et al. [310].

Interdependencies within systems like industrial cool boxes [517] and gearbox systems [518] have been modeled by a dependency matrix to express the interaction of the components in a multicomponent system [519]. This has not yet been done for batteries, but it seems possible given the construction plan of a battery system.

Composing a battery pack of cells, i.e., omitting module assembly, is also called cell-to-pack (CTP) and can reduce parts and increase pack volume utilization [520; 521]. The application of the method proposed in this thesis to BEVs with CTP technology seems possible. Though, it needs to be assumed that aging characteristics change because of a different cooling system and system architecture.

In Chapter 7, a model trained on battery cell data from laboratory operation was transferred to a new domain specified by different laboratory battery operation or another battery type. As suggested in previous work [3] and based on the results from Chapters 7 and 8, a transfer with a double domain shift is thinkable: A model pretrained on battery cell data from laboratory operation could be transferred to the domain of vehicle operation with a battery system composed of the same battery cell. Such an approach could take advantage of more data becoming available in the process from cell and module development and testing to BEV series production as structured by [522]. Especially, cell data from tests on the C- and D-sample stage seem suitable because of their quantity and similarity to series production [510, p. 11].

Then, possibly another advantage of transfer learning might become relevant if models are trained on thousands of samples from battery cells and systems operated BEV fleets. Like in computer vision, then transfer learning might save usage of computational resources when pretraining a model requires weeks of training, but transfer learning significantly less.

10.4 Further Learning Paradigms

Compared to the application of transfer learning in computer vision, pretrained models related to battery aging or SOH forecasting do not exist. This is potentially due to the fact that there are only few publicly available battery aging data sets. Furthermore, in the field of battery aging standardized tasks like RUL prediction and SOH estimation exist, just like image classification and image segmentation in computer vision. However, SotA models focus on feature engineering rather than on the architecture of method or ML model. Thus, the used input features are very different regarding complexity and aggregation which complicates providing pretrained models. Defining a feature set and task with a specified output would improve comparability of SotA models.

Continual Learning

Chapter 7 examined only how the quantity of samples influenced the transfer. Going further, the sequence of training samples from battery operation becoming available can be taken into account. Especially once new data become available, investigations regarding the sequence in which training samples from different domains are used for model training seem relevant. Furthermore, given the data used in this thesis, only a single battery type was considered each in the source and target domain. Future work could examine SOH forecasting for several battery types whose data become available sequentially.

For example, a focus could lie on learning to forecast the SOH of new battery types and different battery operation that has not yet been learned by the model without forgetting the already learned. This concept is called continual learning [523; 524]. Continual learning in the context of SOH forecasting is imaginable in two scenarios: First, with progressing time more and more data of the same battery type or fleet will be available, for instance every day, week, or month. Each time, integrating this new data into an existing model is a task of continual learning. Second, a single SOH forecasting model may be desirable for several battery types of different generations that are only developed in temporal sequence. Once, data of a new battery type is accessible an existing model shall be made suitable for the old battery types and the new one.

Connected to continual learning, curricular learning is worth examining to find a training strategy that presents training samples not randomly, but organized in a meaningful order. This enables the model to gradually learn more complex concepts [525].

Federated Learning

In this thesis, aspects of data privacy protection have not been considered. If data obtained from experimental or test vehicles is used for training an ML model, this may be a smaller concern. However, it will also be desirable to use data from a fleet of BEV owned and operated by private individuals to enlarge the training data set. Then privacy concerns might arise if the data enables conclusions to when, where, and how the BEV were operated.

This concern in the context of fleets and this thesis may be solved by the learning paradigm of federated learning. Federated learning aims to train a central model while the training data stays distributed locally at each client which are the BEVs in this thesis [497]. If the features aggregating the BEV operation are considered not to be GDPR-conform, the data may stay on the BEVs for local training which improves data security and saves costs for central training. Only the local updates of the model parameters are transferred to a central back-end where the updates received from all BEVs are aggregated so that the central model can be updated. Finally, the updated central model is send to all BEVs. Thus, the training data itself never leaves the BEVs which also reduces the cost of data transfer and improves data security. Future challenges for federated learning applied to SOH forecasting are that updates of the model parameters will likely be obtained from different operational domains and different aging phases. This complicates the central aggregation of updates of the model parameters. A possible solution is to apply an unsupervised ML method to cluster the updates of the model parameters which reflect what the current central model has not learned yet. Thereby, groups, anomalies, and outliers may be detected and weighted accordingly in the aggregation so that the model also learns special load patterns and aging behavior [15].

Explainable AI

When providing an SOH forecasting model to end users such as fleet managers, private BEV owners, or battery designers, these users need to trust the model's forecast once they apply it. Trust can be build through explanations and interpretability of the model and its forecasts [526]. However, the inherent black-box character of the MLP applied in this thesis limits explainability and interpretability: When the model is run to forecast, only the model's input and the model's prediction are known—without any explanation. In the context of this thesis, two strategies were followed in further work [4]:

First, as ML models with more inherent white box character DT, RF, XGBoost, and HistXGBoost were used. However, on the same data these models performed worse than ResMLP and two-dimensional ResCNN as DL models. Therefore, the presented DL models seem more suitable than the presented ML models for SOH forecasting even though these two DL models lack inherent white box character.

Thus, a second strategy was applied to these DL models: Gaussian-filtered gradient-based saliency maps were applied to provide post-model interpretability. The saliency maps marked feature regions, i.e., regions of operational states described by the histograms, important to the model's forecast. Successful validation of the saliency maps was conducted by comparison of domain expert knowledge from controlled laboratory aging experiments with those operational states identified as harmful for the battery.

Future research may apply attention-based models to SOH forecasting. Once attention-based models reach a similar performance as models presented in this thesis and by further work [4] in the context of this thesis, their attention weights may be beneficial to improve the interpretability of attention-based models because attention weights reflect the flexible focus of an ANN.

11 Conclusion

SOH forecasting of LIBs concerns the influence of battery usage on their SOH which is especially relevant for BEV fleets. Until now, applicability of SOH forecasting methods to BEV operation has been limited due to a) smaller operational variability from laboratory operation compared to real-world vehicle operation, b) assuming the same load during whole battery life, c) a focus of methods only on cells, but not on battery systems assembled in BEV, and d) a lack of transferability to different battery domains, e.g., new battery types or new operational domains. These limitations also became apparent in a structured literature survey (Chapter 4). To overcome these limitations, the following RQs were investigated:

RQ.1 How can the SOH of LIB cells in laboratory operation be forecasted? (Cell & Lab prognosis, Chapter 6)
In the examined method, through a stressor extraction and a sliding window approach battery operational time series data was sampled with the window width defining the forecast horizon. Thereby, 1D, 2D, and 3D histograms were created that encoded variable battery operation successfully so that an MLP was able to learn forecasting the cell SOH. Using windows of different widths improved the generalization.

RQ.2 How can a model for SOH forecasting be transferred to LIBs with a different cell chemistry or different operational load? (Transfer, Chapter 7)
Based on the previous RQ, this thesis investigated the model transfer to various new cell chemistries and new operational domains. A parametric transfer of a pretrained model was conducted by layerwise freezing. The transfer especially worked well, when at least data of few aged cells was

F. von Bülow, *A Data-Driven Fleet Service: State of Health Forecasting of Lithium-Ion Batteries*, AutoUni – Schriftenreihe 170,
https://doi.org/10.1007/978-3-658-43188-4_11

available. This is preferable over data of many young battery cells as some reference points of aging and operational load of the new cells are required. Contrarily to computer vision, freezing no layers performed better in 95 % of the experimental scenarios.

RQ.3 In what way is a method for SOH forecasting of LIB cells in laboratory operation suitable to forecast the SOH of a LIB system in vehicle operation? (System & BEV prognosis, Chapter 8)
The method for SOH forecasting examined on cell data from laboratory operation was also found to be suitable to forecast the SOH of battery system data from BEV fleet operation. Especially, the operational variability of real-world fleet operation could novelly be captured by the features. Histograms-based features accomplished better results than simpler, more accessible features. However, the overall performance on fleet data was limited by noisy data compared to the SOH forecast of cells and little aging of the batteries.

RQ.4 How can SOH forecasting of LIBs in automotive applications technically be realized? (Open RQ, Chapter 9)
To obtain the training data for the ML model in the proposed SOH forecasting method, a layer-wise concept for a fleet management considering the perspectives of not only the drivers, fleet managers, but also the BEV manufacturer was proposed. This concept also facilitates the provision of an SOH forecasting method as cloud-based fleet service.

In particular, this thesis focused on an SOH forecasting method that is applicable not only in the laboratory, as many other SotA methods, but first and foremost in BEV fleets. Also, the benefit of the method and its ease of application for users like fleet managers was included from the beginning. The model could be provided to fleet managers and customers as a cloud-based fleet service, e.g., in a smartphone or web app. Further, it is possible to run scenario-based interferences with the presented SOH forecasting method to improve the battery operational strategy of a BEV fleet. One exemplarily scenario was defined by a limitation of the SOE while parking (Section 8.4).

The presented methods were able to benefit from data sets both from laboratory and BEV fleet operation. However, the method's performance was limited by the quality of the training data sets, e.g., due to noise, resolution, and measurement accuracy of the used time series signals. Ideas for further research on SOH forecasting laid the foundation for further progress regarding improvement of the method, future data, and other ML learning paradigms. Particularly, the application to battery systems, also with the combination of data on cell-level, is expected to offer potential for further enhancement. Among other aspects, future research may be directed towards a method with a seq2seq input-output-structure to include path dependence on the input side and to directly output an SOH trajectory.

A publicly available data set not from laboratory battery operation, but from real-world vehicle operation with standardization of metric, output value, and forecast horizon would enable benchmarking of SOH forecasting models applicable in real-world vehicle operation. Thus, it would accelerate progress in the research field.

Bibliography

1. Friedrich von Bülow, Joshua Mentz, and Tobias Meisen. "State of health forecasting of Lithium-ion batteries applicable in real-world operational conditions." In: *Journal of Energy Storage* 44 (2021), p. 103439. ISSN: 2352152X. https://doi.org/10.1016/j.est.2021.103439.
2. Friedrich von Bülow and Tobias Meisen. "A review on methods for state of health forecasting of lithium-ion batteries applicable in real-world operational conditions." In: *Journal of Energy Storage* 57 (2023), p. 105978. ISSN: 2352152X. https://doi.org/10.1016/j.est.2022.105978.
3. Friedrich von Bülow, Markus Wassermann, and Tobias Meisen. "State of health forecasting of Lithium-ion batteries operated in a battery electric vehicle fleet." In: *Journal of Energy Storage* 72PA (2023), p. 108271. ISSN: 2352152X. https://doi.org/10.1016/j.est.2023.108271.
4. Friedrich von Bülow, Yannik Hahn, Richard Meyes, and Tobias Meisen. "Transparent and Interpretable State of Health Forecasting of Lithium-ion Batteries with Deep Learning and Saliency Maps." In: *International Journal of Energy Research* 2023 (2023), pp. 1–23. issn: 1099–114X. https://doi.org/10.1155/2023/9922475.
5. Friedrich von Bülow and Tobias Meisen. *State of Health Estimation of Lithium-Ion Batteries under Real-World Operational Conditions. Advanced Battery Power Conference.* 2021. url: https://2021.battery-power.eu/poster2/512/ (visited on 01.03.2023).
6. Friedrich von Bülow, Felix Heinrich, and Tobias Meisen. "Fleet Management Approach for Manufacturers displayed at the Use Case of Battery Electric Vehicles." In: *IEEE International Conference on Systems, Man, and Cybernetics* (SMC) (Melbourne, Australia, 17.10.–20.10.2021). IEEE, 2021, pp. 3218–3225. ISBN: 9781665442084. https://doi.org/10.1109/SMC52423.2021.9658680.
7. Friedrich von Bülow, Felix Heinrich, and Tobias Meisen. "A new perspective for manufacturers: Cloud-based utilization of operational fleet data." In: *ELIV 2021* (Bonn, Germany, 20.10.–21.10.2021). Ed. by VDI Wissensforum GmbH. VDI-Berichte 2384. Düsseldorf, Germany: VDI Verlag, 2021, pp. 117–130. ISBN: 978-3-18-092384-0. https://doi.org/10.51202/9783181023846-117.
8. Friedrich von Bülow and Tobias Meisen. "State of Health Forecasting of Heterogeneous Lithium-ion Battery Types and Operation Enabled by Transfer Learning." In: *PHM Society European Conference* 7 (1 2022), pp. 490–508. https://doi.org/10.36001/phme.2022.v7i1.3312.

F. von Bülow, *A Data-Driven Fleet Service: State of Health Forecasting of Lithium-Ion Batteries*, AutoUni – Schriftenreihe 170,
https://doi.org/10.1007/978-3-658-43188-4_9

9. Friedrich von Bülow, Markus Wassermann, and Tobias Meisen. *State of Health Forecasting of Lithium-Ion Batteries operated in a Mobility on Demand Fleet.* Poster, Batterieforum Deutschland 2023, Berlin, Germany. 2023. url: https://www.batterieforum-deutschland.de/kongress/.
10. Friedrich von Bülow. "Lithium-Ion Battery State of Health Forecasting with Machine Learning and Vehicle Fleet Data." In: *Advanced Automotive Batteries Conference Europe* (AABC) (Mainz, Germany, 19.06.–22.06.2023). 2023. url: https://www.advancedautobat.com/23/aabc-europe/battery-intelligence.
11. Friedrich von Bülow and Felix Heinrich. "Betrieb eines elektrischen Energiespeichers in einem Kraftfahrzeug." G07C 5/08. DE 10 2021 204 847 A1 (Germany). Volkswagen AG. 01.12.2022.
12. Friedrich von Bülow and Felix Heinrich. "Prädiktion einer Charakteristik einer Zielflotte." G07C 5/08. DE 10 2021 204 849 A1 (Germany). Volkswagen AG. 17.11.2022.
13. Friedrich von Bülow, Jonas Krog, and Lukas Kille. "Abwärmenutzung einer Schnelllade-Infrastruktur." B60L 53/302. DE 10 2022 201 609 A1 (Germany). Volkswagen AG. 17.08.2023.
14. Friedrich von Bülow, Fabian Kai-Dietrich Noering, and Daniel Fries. "Verfahren zum Bewerten eines technischen Zustands eines Kraftfahrzeugs mittels eines Bewertungssystems, Computerprogrammprodukt sowie Bewertungssystem." DE 10 2022 213 940.2 (Germany). Volkswagen AG. Publication in process.
15. Andreas Udo Sass and Friedrich von Bülow. "Verfahren und Vorrichtung zum maschinellen Lernen für die Alterungsvorhersage einer Komponente eines Fahrzeugs." DE 10 2023 201 255.3 (Germany). Volkswagen AG. Publication in process.
16. United Nations. *Paris Agreement.* 2015. url: http://unfccc.int/files/essential_background/convention/application/pdf/english_paris_agreement.pdf (visited on 01.03.2023).
17. European Commission. *The European Green Deal. COM(2019) 640 final.* 2019. url: https://eur-lex.europa.eu/legal-content/EN/TXT/?uri=CELEX:52019DC0640 (visited on 01.03.2023).
18. European Environment Agency. *Decarbonising road transport—the role of vehicles, fuels and transport demand. EEA Report No 2/2022.* Ed. by Publications Office of the European Union. 2022. https://doi.org/10.2800/68902. (visited on 01.03.2023).
19. International Energy Agency. *Global Electric Vehicle Outlook 2023. Catching up with climate ambitions.* 2023. url: https://iea.blob.core.windows.net/assets/dacf14d2-eabc-498a-8263-9f97fd5dc327/GEVO2023.pdf (visited on 27.04.2023).
20. ROLAND BERGER GmbH and fka GmbH. *E-Mobility Index 2021.* 2021. url: https://www.rolandberger.com/publications/publication_pdf/roland_berger_e_mobility_index_2021_en.pdf (visited on 01.03.2023).
21. Forbes Media LLC. *Every Automaker's EV Plans Through 2035 And Beyond.* 2021. url: https://www.forbes.com/wheels/news/automaker-ev-plans/ (visited on 01.03.2023).
22. Koki Ogura and Mohan Lal Kolhe. "Battery technologies for electric vehicles." In: *Electric Vehicles: Prospects and Challenges.* Ed. by Tariq Muneer, Mohan Lal Kolhe, and Aisling Doyle. 1st ed. Amsterdam, The Netherlands: Elsevier, 2017, pp. 139–167. ISBN: 9780128030219. https://doi.org/10.1016/B978-0-12-803021-9.00004-5.
23. Yuanli Ding, Zachary P. Cano, Aiping Yu, Jun Lu, and Zhongwei Chen. "Automotive Li-Ion Batteries. Current Status and Future Perspectives." In: *Electrochemical Energy*

Reviews 2 (1 2019), pp. 1–28. ISSN: 2520-8489. https://doi.org/10.1007/s41918-018-0022-z.

24. Philipp André Dechent, Alexander Epp, Dominik Jöst, Yuliya Preger, Peter M. Attia, Weihan Li, and Dirk Uwe Sauer. "ENPOLITE. Comparing Lithium-Ion Cells across Energy, Power, Lifetime, and Temperature." In: ACS Energy Letters 6 (6 2021), pp. 2351–2355. ISSN: 2380-8195. https://doi.org/10.1021/acsenergylett.1c00743.
25. Stephan Leuthner. "Lithium-ion battery overview." In: *Lithium-Ion Batteries: Basics and Applications*. Ed. by Reiner Korthauer. Berlin, Heidelberg, Germany: Springer Berlin Heidelberg, 2018, pp. 13–19. ISBN: 978-3-662-53069-6. https://doi.org/10.1007/978-3-662-53071-9_2.
26. Kai-Christian Moeller. "Overview of battery systems." In: *Lithium-Ion Batteries: Basics and Applications*. Ed. by Reiner Korthauer. Berlin, Heidelberg, Germany: Springer Berlin Heidelberg, 2018, pp. 3–9. ISBN: 978-3-662-53069-6. https://doi.org/10.1007/978-3-662-53071-9_1.
27. Gianfranco Pistoia. "APPLICATIONS—PORTABLE | Portable Devices: Batteries." In: *Encyclopedia of electrochemical power sources*. Ed. by Jürgen Garche. 1st ed. Amsterdam, The Netherlands: Elsevier, 2009, pp. 29–38. ISBN: 978-0-444-52745-5. https://doi.org/10.1016/B978-044452745-5.00358-0.
28. Ralph J. Brodd. "APPLICATIONS—PORTABLE | Notebooks: Batteries." In: *Encyclopedia of electrochemical power sources*. Ed. by Jürgen Garche. 1st ed. Amsterdam, The Netherlands: Elsevier, 2009, pp. 22–28. ISBN: 978-0-444-52745-5. https://doi.org/10.1016/B978-044452745-5.00359-2.
29. Alvaro Masias, James Marcicki, and William A. Paxton. "Opportunities and Challenges of Lithium Ion Batteries in Automotive Applications." In: *ACS Energy Letters* 6 (2 2021), pp. 621–630. ISSN: 2380-8195. https://doi.org/10.1021/acsenergylett.0c02584.
30. Languang Lu, Xuebing Han, Jianqiu Li, Jianfeng Hua, and Minggao Ouyang. "A review on the key issues for lithium-ion battery management in electric vehicles." In: *Journal of Power Sources* 226 (2013), pp. 272–288. ISSN: 03787753. https://doi.org/10.1016/j.jpowsour.2012.10.060.
31. StartUs Insights. *5 Top Battery Analytics Startups*. 2022. url: https://www.startusinsights.com/innovators-guide/battery-analytics-startups/ (visited on 01.03.2023).
32. Adrian König, Lorenzo Nicoletti, Daniel Schröder, Sebastian Wolff, Adam Waclaw, and Markus Lienkamp. "An Overview of Parameter and Cost for Battery Electric Vehicles." In: *World Electric Vehicle Journal* 12 (1 2021), p. 21. ISSN: 2032-6653. https://doi.org/10.3390/wevj12010021.
33. United States Advanced Battery Consortium. *Goals for Advanced Batteries for EVs. CY 2020 Commercialization*. 2019. url: https://uscar.org/download/246/energystorage-system-goals/12837/goals-for-advanced-high-performance-batteries-forelectric-vehicle-ev-applications.pdf (visited on 01.03.2023).
34. Alvaro Masias. "Lithium-Ion Battery Design for Transportation." In: *Behaviour of Lithium-Ion Batteries in Electric Vehicles. Battery Health, Performance, Safety, and Cost*. Ed. by Boryann Liaw and Gianfranco Pistoia. 1st ed. Green Energy and Technology. Cham, Switzerland: Springer International Publishing and Springer, 2018, pp. 1–33. ISBN: 9783319699509. https://doi.org/10.1007/978-3-319-69950-9_1.

35. Mahammad Abdul Hannan, Molla Shahadat Hossain Lipu, Aini Hussain, and Azah M. Mohamed. "A review of lithium-ion battery state of charge estimation and management system in electric vehicle applications. Challenges and recommendations." In: *Renewable and Sustainable Energy Reviews* 78 (9 2017), pp. 834–854. ISSN: 13640321. https://doi.org/10.1016/j.rser.2017.05.001.
36. Florian Herrmann and Florian Rothfuss. "Introduction to hybrid electric vehicles, battery electric vehicles, and off-road electric vehicles." In: *Advances in battery technologies for electric vehicles.* Ed. by Bruno Scrosati, Jürgen Garche, and Werner Tillmetz. 1st ed.Woodhead Publishing series in energy 80. Cambridge, UK:Woodhead Publishing, 2015, pp. 3–16. ISBN: 9781782423775. https://doi.org/10.1016/B978-1-78242-377-5.00001-7.
37. Dirk Magnor, Benedikt Lunz, and Dirk Uwe Sauer. "'Double Use' of Storage Systems." In: *Electrochemical energy storage for renewable sources and grid balancing.* Ed. by Jürgen Garche, Patrick T. Moseley, and Peter Adelmann. Amsterdam, The Netherlands: Elsevier, 2015, pp. 453–463. ISBN: 9780444626165. https://doi.org/10.1016/B978-0-444-62616-5.00022-X.
38. Jones Miller. *Electric car costs to remain higher than traditional engines.* Ed. By THE FINANCIAL TIMES LTD. 2020. url: https://www.ft.com/content/a7e58ce7-4fab-424a-b1fa-f833ce948cb7 (visited on 01.03.2023).
39. Song Su-hyun. *[Chew on I.T.] What makes up smartphone prices?* Ed. by The Korea Herald by Herald Corporation. 2020. url: https://www.koreaherald.com/view.php?ud=20200607000076 (visited on 01.03.2023).
40. Al Cowsky. *Cost Comparison—Huawei Mate 10, iPhone 8, Samsung Galaxy S8.* Ed. by TechInsights Inc. 2017. url: https://www.techinsights.com/blog/costcomparison-huawei-mate-10-iphone-8-samsung-galaxy-s8 (visited on 01.03.2023).
41. Susan Shaheen and Adam Cohen. "Similarities and Differences of Mobility on Demand (MOD) and Mobility as a Service (MaaS)." In: *Institute of Transportation Engineers Journal* 90 (6 2020), pp. 29–35. ISSN: 0162-8178.
42. Susan Shaheen and Adam Cohen. "Chapter 3—Mobility on demand (MOD) and mobility as a service (MaaS): early understanding of shared mobility impacts and public transit partnerships." In: *Demand for Emerging Transportation Systems.* Ed. by Dimitrios Efthymiou, Constantinos Antoniou, and Emmanouil Chaniotakis. Elsevier, 2020, pp. 37–59. ISBN: 978-0-12-815018-4. https://doi.org/10.1016/B978-0-12-815018-4.00003-6.
43. Cláudia Machado, Nicolas de Salles Hue, Fernando Berssaneti, and José Quintanilha. "An Overview of Shared Mobility." In: *Sustainability* 10 (12 2018), p. 4342. ISSN: 2071-1050. https://doi.org/10.3390/su10124342.
44. UMI Urban Mobility. *WeShare 100% elektrisches Carsharing.* url: https://www.weshare.io/ (visited on 01.03.2023).
45. Quantum Lifecycle Partners. *What's the Average Lifespan of Your Electronics?* 2021. url: https://quantumlifecycle.com/en_CA/blog/whats-the-average-lifespan-of-yourelectronics/ (visited on 01.03.2023).
46. European Council for Automotive R&D. *Battery requirements for future automotive applications.* 2019. url: https://eucar.be/wp-content/uploads/2019/08/20190710-EG-BEV-FCEV-Battery-requirements-FINAL.pdf (visited on 01.03.2023).
47. New Energy and Industrial Technology Development Organization. "The Japanese policy and NEDO activity for future mobility." In: (2017). url: https://www.tresor.

economie.gouv.fr/Articles/70dfe615-29bf-4ee4-a232-2fd2b954c025/files/59a6914a-57e1-495f-8108-abdecb7d0715 (visited on 01.03.2023).

48. Jeremy Neubauer and Eric Wood. "The impact of range anxiety and home, workplace, and public charging infrastructure on simulated battery electric vehicle lifetime utility." In: *Journal of Power Sources* 257 (2014), pp. 12–20. ISSN: 03787753. https://doi.org/10.1016/j.jpowsour.2014.01.075.
49. John Warner. *The Handbook of Lithium-Ion Battery Pack Design. Chemistry, components, types and terminology.* Amsterdam, The Netherlands: Elsevier, 2015. 263 pp. ISBN: 9780128016688. https://doi.org/10.1016/C2013-0-23144-5.
50. Roland Dorn, Reiner Schwartz, and Bjoern Steurich. "Battery management system." In: *Lithium-Ion Batteries: Basics and Applications.* Ed. by Reiner Korthauer. Berlin, Heidelberg, Germany: Springer Berlin Heidelberg, 2018, pp. 165–175. ISBN: 978-3-662-53069-6. https://doi.org/10.1007/978-3-662-53071-9_14.
51. Wladislaw Waag and Dirk Uwe Sauer. "SECONDARY BATTERIES—LEAD-ACID SYSTEMS | State-of-Charge/Health." In: *Encyclopedia of electrochemical power sources.* Ed. by Jürgen Garche. 1st ed. Amsterdam, The Netherlands: Elsevier, 2009, pp. 793–804. ISBN: 978-0-444-52745-5. https://doi.org/10.1016/B978-044452745-5.00149-0.
52. Felix Heinrich and Marco Pruckner. "Virtual experiments for battery state of health estimation based on neural networks and in-vehicle data." In: *Journal of Energy Storage* 48 (116 2022), p. 103856. ISSN: 2352152X. https://doi.org/10.1016/j.est.2021.103856.
53. Verena Klass. "Battery Health Estimation." PhD Thesis. Stockholm, Sweden, 2015. 71 pp. url: https://www.diva-portal.org/smash/get/diva2:853550/FULLTEXT01.pdf.
54. Yaxiang Fan, Fei Xiao, Chaoran Li, Guorun Yang, and Xin Tang. "A novel deep learning framework for state of health estimation of lithium-ion battery." In: *Journal of Energy Storage* 32 (2020), p. 101741. ISSN: 2352152X. https://doi.org/10.1016/j.est.2020.101741.
55. Jinsong Yu, Baohua Mo, Diyin Tang, Jie Yang, Jiuqing Wan, and Jingjing Liu. "Indirect state-of-health estimation for lithium-ion batteries under randomized use." In: *Energies* 10 (12 2017), p. 2012. ISSN: 1996-1073. https://doi.org/10.3390/en10122012.
56. Ji Wu, Yujie Wang, Xu Zhang, and Zonghai Chen. "A novel state of health estimation method of li-ion battery using group method of data handling." In: *Journal of Power Sources* 327 (2016), pp. 457–464. ISSN: 03787753. https://doi.org/10.1016/j.jpowsour.2016.07.065.
57. Anthony Barré, Benjamin Deguilhem, Sébastien Grolleau, Mathias Gérard, Frédéric Suard, and Delphine Riu. "A review on lithium-ion battery ageing mechanisms and estimations for automotive applications." In: *Journal of Power Sources* 241 (9 2013), pp. 680–689. ISSN: 03787753. https://doi.org/10.1016/j.jpowsour.2013.05.040.
58. Tanja Gewald, Adrian Candussio, Leo Wildfeuer, Dirk Lehmkuhl, Alexander Hahn, and Markus Lienkamp. "Accelerated aging characterization of lithium-ion cells. Using sensitivity analysis to identify the stress factors relevant to cyclic aging." In: *Batteries* 6 (1 2020), p. 6. ISSN: 2313–0105. https://doi.org/10.3390/batteries6010006.
59. Madeleine Ecker, Nerea Nieto, Stefan Käbitz, Johannes Schmalstieg, Holger Blanke, Alexander Warnecke, and Dirk Uwe Sauer. "Calendar and cycle life study of Li(NiMnCo)O2-based 18650 lithium-ion batteries." In: *Journal of Power Sources* 248

(9 2014), pp. 839–851. ISSN: 03787753. https://doi.org/10.1016/j.jpowsour.2013.09.143.

60. Lérys Granado, Mohamed Ben-Marzouk, Eduard Solano Saenz, Yassine Boukal, and Sylvain Jugé. "Machine learning predictions of lithium-ion battery state-of-health for eVTOL applications." In: *Journal of Power Sources* 548 (2 2022), p. 232051. ISSN: 03787753. https://doi.org/10.1016/j.jpowsour.2022.232051.
61. Heinz Wenzl. "BATTERIES AND FUEL CELLS | Lifetime." In: *Encyclopedia of electrochemical power sources*. Ed. by Jürgen Garche. 1st ed. Amsterdam, The Netherlands: Elsevier, 2009, pp. 552–558. ISBN: 978-0-444-52745-5. https://doi.org/10.1016/B978-044452745-5.00048-4.
62. Nikolaos Wassiliadis, Jakob Schneider, Alexander Frank, Leo Wildfeuer, Xue Lin, Andreas Jossen, and Markus Lienkamp. "Review of fast charging strategies for lithium-ion battery systems and their applicability for battery electric vehicles." In: *Journal of Energy Storage* 44 (6 2021), p. 103306. ISSN: 2352152X. https://doi.org/10.1016/j.est.2021.103306.
63. Bogdan-Adrian Enache, George-Calin Seritan, Sorin-Dan Grigorescu, Costin Cepisca, Felix-Constantin Adochiei, Violeta-Vasilica Argatu, and Teodor Iulian Voicila. "A Battery Screening System for Second Life LiFePO4 Batteries." In: *International Conference and Exposition on Electrical And Power Engineering (EPE)*. 2020 International Conference and Exposition on Electrical And Power Engineering (EPE) (Iasi, Romania, 22.10.–23.10.2020). IEEE, 2020, pp. 298–301. ISBN: 978-1-7281-8126-4. https://doi.org/10.1109/EPE50722.2020.9305538.
64. Zihao Zhou, Aihua Ran, Shuxiao Chen, Xuan Zhang, Guodan Wei, Baohua Li, Feiyu Kang, Xiang Zhou, and Hongbin Sun. "A fast screening framework for second-life batteries based on an improved bisecting K-means algorithm combined with fast pulse test." In: *Journal of Energy Storage* 31 (3 2020), p. 101739. ISSN: 2352152X. https://doi.org/10.1016/j.est.2020.101739.
65. Münchener Rückversicherungs-Gesellschaft. *Battery performance now insurable—Innovative Munich Re coverage paves the way for renewable energy.* 2019. url: https://www.munichre.com/content/dam/munichre/contentlounge/website-pieces/documents/Munich-Re-PM-2019-03-07_en.pdf (visited on 01.03.2023).
66. Ayca Zeynep Suer. "The Recognition of Provisions. Evidence from BIST100 Nonfinancial Companies." In: *Procedia Economics and Finance* 9 (2014), pp. 391–401. ISSN: 22125671. https://doi.org/10.1016/S2212-5671(14)00040-9.
67. Deutscher Bundestag. *Handelsgesetzbuch*. HGB. 10.03.2023. url: https://www.gesetze-im-internet.de/englisch_hgb/englisch_hgb.html#p0802 (visited on 01.03.2023).
68. Brian Bole, Chetan S. Kulkarni, and Matthew Daigle. *Randomized Battery Usage Data Set.* NASA Ames Prognostics Data Repository, NASA Ames Research Center, Moffett Field, CA. 2012. url: https://www.nasa.gov/content/prognostics-center-ofexcellence-data-set-repository (visited on 01.03.2023).
69. Peter M. Attia, Aditya Grover, Norman Jin, Kristen A. Severson, Todor M. Markov, Yang-Hung Liao, Michael H. Chen, Bryan Cheong, Nicholas Perkins, Zi Yang, Patrick K. Herring, Muratahan Aykol, Stephen J. Harris, Richard D. Braatz, Stefano Ermon, and William C. Chueh. "Closed-loop optimization of fast-charging protocols for batteries with machine learning." In: *Nature* 578 (7795 2020), pp. 397–402. https://doi.org/10.1038/s41586-020-1994-5.

70. Kristen A. Severson, Peter M. Attia, Norman Jin, Nicholas Perkins, Benben Jiang, Zi Yang, Michael H. Chen, Muratahan Aykol, Patrick K. Herring, Dimitrios Fraggedakis, Martin Z. Bazant, Stephen J. Harris, William C. Chueh, and Richard D. Braatz. "Data-driven prediction of battery cycle life before capacity degradation." In: *Nature Energy* 4 (5 2019), pp. 383–391. ISSN: 2058-7546. https://doi.org/10.1038/s41560-019-0356-8.
71. Bhaskar Saha and Kai Goebel. *Battery Data Set*. NASA Ames Prognostics Data Repository, NASA Ames Research Center, Moffett Field, CA. 2008. url: https://www.nasa.gov/content/prognostics-center-of-excellence-data-set-repository (visited on 01.03.2023).
72. Yunhong Che, Xiaosong Hu, Xianke Lin, Jia Guo, and Remus Teodorescu. "Health prognostics for lithium-ion batteries. Mechanisms, methods, and prospects." In: *Energy & Environmental Science* 16 (2 2023), pp. 338–371. ISSN: 1754-5692. https://doi.org/10.1039/d2ee03019e.
73. Shaheer Ansari, Afida Ayob, Molla Shahadat Hossain Lipu, Aini Hussain, and Mohamad Hanif Md Saad. "Remaining useful life prediction for lithium-ion battery storage system. A comprehensive review of methods, key factors, issues and future outlook." In: *Energy Reports* 8 (2022), pp. 12153–12185. ISSN: 2352-4847. https://doi.org/10.1016/j.egyr.2022.09.043.
74. Xiaohong Wang, Zhuo Wang, Lizhi Wang, Zhuoqi Wang, and Hongzhou Guo. "Dependency analysis and degradation process-dependent modeling of lithium-ion battery packs." In: *Journal of Power Sources* 414 (2019), pp. 318–326. ISSN: 03787753. https://doi.org/10.1016/j.jpowsour.2019.01.021.
75. Heinz Wenzl, André Haubrock, and Hans-Peter Beck. "Degradation of Lithium Ion Batteries under Complex Conditions of Use." In: *Zeitschrift für Physikalische Chemie* 227 (1 2013), pp. 57–72. ISSN: 2196-7156. https://doi.org/10.1524/zpch.2012.0170.
76. Luc Pelkmans, Staffan Hultén, Gotzon Arkarate, and Panayotis Christidis. "Trends in vehicle and fuel technologies: Review of past trends." In: *Institute for Prospective Technological Studies Technical Report Series, EUR 20746 EN* (2003). url: http://temis.documentation.developpement-durable.gouv.fr/docs/Temis/0073/Temis-0073613/21152.pdf (visited on 01.03.2023).
77. Roger Young, Sheila Fallon, Paul Jacob, and Denis O'Dwyer. "Vehicle Telematics and Its Role as a Key Enabler in the Development of Smart Cities." In: *IEEE Sensors Journal* 20 (19 2020), pp. 11713–11724. https://doi.org/10.1109/JSEN.2020.2997129.
78. Christopher Vox, David Broneske, Jan Piewek, Andreas Udo Sass, and Gunter Saake. "Integer Time Series Compression for Holistic Data Analytics in the Context ofVehicle Sensor Data." In: *International Conference on Connected Vehicle and Expo (ICCVE)* (Lakeland, FL, US, 07.03.–09.03.2022). IEEE, 2022. ISBN: 978-1-6654-1687-0. https://doi.org/10.1109/ICCVE52871.2022.9743019.
79. William Fleming. "Forty-Year Review of Automotive Electronics. A Unique Source of Historical Information on Automotive Electronics." In: *IEEE Vehicular Technology Magazine* 10 (3 2015), pp. 80–90. ISSN: 1556-6072. https://doi.org/10.1109/MVT.2015.2447851.
80. Fabio Giust, Vincenzo Sciancalepore, Dario Sabella, Miltiades C. Filippou, Simone Mangiante, Walter Featherstone, and Daniele Munaretto. "Multi-Access Edge Computing. The Driver Behind the Wheel of 5G-Connected Cars." In: *IEEE Communications*

Standards Magazine 2 (3 2018), pp. 66–73. ISSN: 2471-2825. https://doi.org/10.1109/MCOMSTD.2018.1800013.

81. Zaigham Mahmood. "Connected Vehicles in the IoV: Concepts, Technologies and Architectures." In: *Connected Vehicles in the Internet of Things. Concepts, Technologies and Frameworks for the IoV.* Ed. by Zaigham Mahmood. 1st ed. Cham, Switzerland: Springer International Publishing, 2020, pp. 3–18. ISBN: 978-3-030-36167-9.
82. David Gomez-Barquero, Wei Li, Manuel Fuentes, Jian Xiong, Giuseppe Araniti, Cristiano Akamine, and Jintao Wang. "IEEE Transactions on Broadcasting Special Issue on. 5G for Broadband Multimedia Systems and Broadcasting." In: *IEEE Transactions on Broadcasting* 65 (2 2019), pp. 351–355. ISSN: 0018-9316. https://doi.org/10.1109/TBC.2019.2914866.
83. Patrick Killeen, Bo Ding, Iluju Kiringa, and Tet Yeap. "IoT-based predictive maintenance for fleet management." In: *Procedia Computer Science* 151 (2019), pp. 607–613. ISSN: 18770509. https://doi.org/10.1016/j.procs.2019.04.184.
84. Uwe Kubach. "Device Clouds: Cloud-Plattformen schlagen die Brücke zwischen Industrie 4.0 und dem Internet der Dinge." In: *Handbuch Industrie 4.0 Bd. 3. Logistik.* Ed. by Birgit Vogel-Heuser, Thomas Bauernhansl, and Michael ten Hompel. 1st ed. Berlin, Germany: Springer, 2017, pp. 181–200. ISBN: 978-3-662-53250-8. https://doi.org/10.1007/978-3-662-53251-5_14.
85. Michael Kleinemeier. "Von der Automatisierungspyramide zu Unternehmenssteuerungsnetzwerken." In: *Industrie 4.0 in Produktion, Automatisierung und Logistik.* Ed. by Thomas Bauernhansl, Michael ten Hompel, and Birgit Vogel-Heuser. 1st ed. Wiesbaden, Germany: Springer Fachmedien Wiesbaden, 2014, pp. 571–579. ISBN: 978-3-658-04681-1. https://doi.org/10.1007/978-3-658-04682-8_29.
86. McKinsey. *The future of mobility is at our doorstep.* 2019. url: https://www.mckinsey.com/industries/automotive-and-assembly/our-insights/the-future-of-mobility-is-atour-doorstep (visited on 01.03.2023).
87. Umar Zakir Abdul Hamid and Fadi Al-Turjman. "Introductory Chapter: A Brief Overview of Autonomous, Connected, Electric and Shared (ACES) Vehicles as the Future of Mobility." In: *Towards Connected and Autonomous Vehicle Highways.* Ed. by Umar Zakir Abdul Hamid and Fadi Al-Turjman. Cham, Switzerland: Springer International Publishing, 2021, pp. 3–8. ISBN: 978-3-030-66042-0. https://doi.org/10.1007/978-3-030-66042-0_1.
88. Boone B. Owens, Priscilla Reale, and Bruno Scrosati. "PRIMARY BATTERIES | Overview." In: *Encyclopedia of electrochemical power sources.* Ed. by Jürgen Garche. 1st ed. Amsterdam, The Netherlands: Elsevier, 2009, pp. 22–27. ISBN: 978-0-444-52745-5. https://doi.org/10.1016/B978-044452745-5.00096-4.
89. Verband der Elektrotechnik Elektronik Informationstechnik e. V. *Battery Systems for Multiple Units. Emission-free drives powered by lithium-ion cells.* 2018. url: https://www.vde.com/resource/blob/2068324/a35ebed9833dae59d8cb1451368203bd/vdestudy-battery-systems-data.pdf (visited on 01.03.2023).
90. Marie-Therese von Srbik. "Advanced lithium-ion battery modelling for automotive applications." Department of Mechanical Engineering. PhD Thesis. London, UK, 2015. url: https://doi.org/10.25560/51498.

91. Peter Keil. "Aging of Lithium-Ion Batteries in Electric Vehicles." Lehrstuhl für Elektrische Energiespeichertechnik. PhD Thesis. München, Germany, 2017. 186 pp. url: https://mediatum.ub.tum.de/doc/1355829/file.pdf.
92. Heff Dahn and Grant M. Ehrlich. "Lithium-Ion Batteries." In: *Linden's handbook of batteries*. Ed. by Thomas B. Reddy and David Linden. 4th ed. New York, NY, US: McGraw-Hill, 2011, pp. 26.1–26.80. ISBN: 9780071624213.
93. Angel Kirchev. "Battery Management and Battery Diagnostics." In: *Electrochemical energy storage for renewable sources and grid balancing*. Ed. by Jürgen Garche, Patrick T. Moseley, and Peter Adelmann. Amsterdam, The Netherlands: Elsevier, 2015, pp. 411–435. ISBN: 9780444626165. https://doi.org/10.1016/B978-0-444-62616-5.00020-6.
94. Hubert Rubenbauer and Stefan Henninger. "Definitions and reference values for battery systems in electrical power grids." In: *Journal of Energy Storage* 12 (2017), pp. 87–107. ISSN: 2352152X. https://doi.org/10.1016/j.est.2017.04.004.
95. Terry Renner. "Tables of physical quantities." In: *Quantities, units and symbols in physical chemistry*. Ed. by E. Richard Cohen, Tom Cvitas, Jeremy G. Frey, Bertil Holström, Kozo Kuchitsu, Roberto Marquardt, Ian Mills, Franco Pavese, Martin Quack, Jüergen Stohner, Herbert L. Strauss, Michio Takami, and Anders J. Thor. 3rd ed. Cambridge: IUPAC & RSC Publishing, 2008, pp. 11–248. ISBN: 978-0-85404-433-7. https://doi.org/10.1039/9781847557889-00011.
96. International Organization for Standardization. ISO/IEC 60050-482:2004. 2004. *International Electrotechnical Vocabulary – Part 482: Primary and secondary cells and batteries*.
97. Anup Barai, Kotub Uddin, Matthieu Dubarry, Limhi Somerville, Andrew McGordon, Paul Jennings, and Ira Bloom. "A comparison of methodologies for the non-invasive characterisation of commercial Li-ion cells." In: *Progress in Energy and Combustion Science* 72 (6058 2019), pp. 1–31. ISSN: 03601285. https://doi.org/10.1016/j.pecs.2019.01.001.
98. Heinz Wenzl. "BATTERIES | Capacity." In: *Encyclopedia of electrochemical power sources*. Ed. by Jürgen Garche. 1st ed. Amsterdam, The Netherlands: Elsevier, 2009, pp. 395–400. ISBN: 978-0-444-52745-5. https://doi.org/10.1016/B978-044452745-5.00043-5.
99. Molla Shahadat Hossain Lipu, Mahammad Abdul Hannan, Aini Hussain, Md. Murshadul Hoque, Pin Jern Ker, Mohamad Hanif Md Saad, and Afida Ayob. "A review of state of health and remaining useful life estimation methods for lithium-ion battery in electric vehicles. Challenges and recommendations." In: *Journal of Cleaner Production* 205 (2018), pp. 115–133. ISSN: 0959-6526. https://doi.org/10.1016/j.jclepro.2018.09.065.
100. Simon F. Schuster, Martin J. Brand, Philipp Berg, Markus Gleissenberger, and Andreas Jossen. "Lithium-ion cell-to-cell variation during battery electric vehicle operation." In: *Journal of Power Sources* 297 (2015), pp. 242–251. ISSN: 03787753. https://doi.org/10.1016/j.jpowsour.2015.08.001.
101. David Linden. "Factors Affecting Battery Performance." In: *Linden's handbook of batteries*. Ed. by Thomas B. Reddy and David Linden. 4th ed. New York, NY, US: McGraw-Hill, 2011, pp. 3.1–3.22. ISBN: 9780071624213.

102. Slobodan Petrovic. *Battery Technology Crash Course.* 1st ed. Cham, Switzerland: Springer International Publishing, 2021. 123 pp. https://doi.org/10.1007/978-3-030-57269-3.
103. Kompetenznetzwerk Lithium-Ionen-Batterien e. V. *C-Koeffizient.* 2022. url: https://www.batterieforum-deutschland.de/infoportal/lexikon/c-koeffizient/ (visited on 01.03.2023).
104. Dirk Uwe Sauer. "BATTERIES I Charge-Discharge Curves." In: *Encyclopedia of electrochemical power sources.* Ed. by Jürgen Garche. 1st ed. Amsterdam, The Netherlands: Elsevier, 2009, pp. 443–451. ISBN: 978-0-444-52745-5. https://doi.org/10.1016/B978-044452745-5.00052-6.
105. MIT Electric Vehicle Team. *A Guide to Understanding Battery Specifications.* 2008. url: http://web.mit.edu/evt/summary_battery_specifications.pdf (visited on 01.03.2023).
106. Kong Soon Ng, Chin-Sien Moo, Yi-Ping Chen, and Yao-Ching Hsieh. "Enhanced coulomb counting method for estimating state-of-charge and state-of-health of lithium-ion batteries." In: *Applied Energy* 86 (9 2009), pp. 1506–1511. ISSN: 03062619; 1872-9118. https://doi.org/10.1016/j.apenergy.2008.11.021.
107. Hongwen He, Yongzhi Zhang, Rui Xiong, and Chun Wang. "A novel Gaussian model based battery state estimation approach. State-of-Energy." In: *Applied Energy* 151 (8 2015), pp. 41–48. ISSN: 1872-9118. https://doi.org/10.1016/j.apenergy.2015.04.062.
108. Eberhard Meissner and Gerolf Richter. "The challenge to the automotive battery industry. The battery has to become an increasingly integrated component within the vehicle electric power system." In: *Journal of Power Sources* 144 (2 2005), pp. 438–460. ISSN: 03787753. https://doi.org/10.1016/j.jpowsour.2004.10.031.
109. Alexander Kohs. *Batteriemodell zur Prädiktion des Gesundheitszustands von Lithium-Ionen-Batterien.* 1st ed. Wissenschaftliche Reihe Fahrzeugtechnik Universität Stuttgart. Wiesbaden, Germany: Springer Fachmedien Wiesbaden, 2022. Online-Ressource. ISBN: 978-3-658-37679-6. https://doi.org/10.1007/978-3-658-37679-6.
110. Andreas Hauser and Robert M. Kuhn. "Cell balancing, battery state estimation, and safety aspects of battery management systems for electric vehicles." In: *Advances in battery technologies for electric vehicles.* Ed. by Bruno Scrosati, Jürgen Garche, and Werner Tillmetz. 1st ed. Woodhead Publishing series in energy 80. Cambridge, UK: Woodhead Publishing, 2015, pp. 283–326. ISBN: 9781782423775. https://doi.org/10.1016/B978-1-78242-377-5.00012-1.
111. Seyed Mohammad Rezvanizaniani, Zongchang Liu, Yan Chen, and Jay Lee. "Review and recent advances in battery health monitoring and prognostics technologies for electric vehicle (EV) safety and mobility." In: *Journal of Power Sources* 256 (2014), pp. 110–124. ISSN: 03787753. https://doi.org/10.1016/j.jpowsour.2014.01.085.
112. Saurabh Saxena, Yinjiao Xing, and Michael G. Pecht. "PHM of Li-ion Batteries." In: *Prognostics and Health Management of Electronics.* Ed. by Michael G. Pecht and Myeongsu Kang. Chichester, UK: JohnWiley and Sons Ltd, 2018, pp. 349–375. ISBN: 9781119515326. https://doi.org/10.1002/9781119515326.ch13.
113. Lin Chen, Zhiqiang Lü, Weilong Lin, Junzi Li, and Haihong Pan. "A new state-of-health estimation method for lithium-ion batteries through the intrinsic relationship between ohmic internal resistance and capacity." In: *Meausrement* 116 (2018), pp. 586–595. ISSN: 02632241. https://doi.org/10.1016/j.measurement.2017.11.016.

114. Xiaosong Hu, Fei Feng, Kailong Liu, Lei Zhang, Jiale Xie, and Bo Liu. "State estimation for advanced battery management. Key challenges and future trends." In: *Renewable and Sustainable Energy Reviews* 114 (2019), p. 109334. ISSN: 13640321. https://doi.org/10.1016/j.rser.2019.109334.
115. Wladislaw Waag, Christian Fleischer, and Dirk Uwe Sauer. "Critical review of the methods for monitoring of lithium-ion batteries in electric and hybrid vehicles." In: *Journal of Power Sources* 258 (2014), pp. 321–339. ISSN: 03787753. https://doi.org/10.1016/j.jpowsour.2014.02.064.
116. Karanjot Singh, Tegoeh Tjahjowidodo, Loïc Boulon, and Mir Feroskhan. "Framework for measurement of battery state-of-health (resistance) integrating overpotential effects and entropy changes using energy equilibrium." In: *Energy* 239 (9 2022), p. 121942. ISSN: 1873-6785. https://doi.org/10.1016/j.energy.2021.121942.
117. Alexander Farmann, Wladislaw Waag, Andrea Marongiu, and Dirk Uwe Sauer. "Critical review of on-board capacity estimation techniques for lithium-ion batteries in electric and hybrid electric vehicles." In: *Journal of Power Sources* 281 (0 2015), pp. 114–130. ISSN: 03787753. https://doi.org/10.1016/j.jpowsour.2015.01.129.
118. Sébastien Grolleau, Arnaud Delaille, Hamid Gualous, Philippe Gyan, Renaud Revel, Julien Bernard, Eduardo Redondo-Iglesias, Jérémy Peter, and SIMCAL Network. "Calendar aging of commercial graphite/LiFePO4 cell—Predicting capacity fade under time dependent storage conditions." In: *Journal of Power Sources* 255 (2014), pp. 450–458. ISSN: 03787753. https://doi.org/10.1016/j.jpowsour.2013.11.098.
119. Clemens Guenther, Benjamin Schott, Wilfried Hennings, Paul Waldowski, and Michael A. Danzer. "Model-based investigation of electric vehicle battery aging by means of vehicle-to-grid scenario simulations." In: *Journal of Power Sources* 239 (2013), pp. 604–610. ISSN: 03787753. https://doi.org/10.1016/j.jpowsour.2013.02.041.
120. Kelli Mamadou, Arnaud Delaille, Elisabeth Lemaire-Potteau, and Yann Bultel. "The State-of-Energy. A New Criterion for the Energetic Performances Evaluation of Electrochemical Storage Devices." In: *ECS Transactions* 25 (35 2010), pp. 105–112. ISSN: 1938-5862. https://doi.org/10.1149/1.3414008.
121. Weiping Diao, Jiuchun Jiang, Caiping Zhang, Hui Liang, and Michael G. Pecht. "Energy state of health estimation for battery packs based on the degradation and inconsistency." In: *Energy Procedia* 142 (9 2017), pp. 3578–3583. ISSN: 18766102. https://doi.org/10.1016/j.egypro.2017.12.248.
122. Nassim Noura, Loïc Boulon, and Samir Jemeï. "A Review of Battery State of Health Estimation Methods. Hybrid Electric Vehicle Challenges." In: *World Electric Vehicle Journal* 11 (4 2020), p. 66. ISSN: 2032-6653. https://doi.org/10.3390/wevj11040066.
123. Mikel Arrinda, Mikel Oyarbide, Haritz Macicior, Eñaut Muxika, Hartmut Popp, Marcus Jahn, Boschidar Ganev, and Iosu Cendoya. "Application Dependent End-of-Life Threshold Definition Methodology for Batteries in Electric Vehicles." In: *Batteries* 7 (1 2021), p. 12. ISSN: 2313-0105. https://doi.org/10.3390/batteries7010012.
124. Yi Li, Kailong Liu, Aoife M. Foley, Alana Zulke, Maitane Berecibar, Elise Nanini-Maury, Joeri van Mierlo, and Harry E. Hoster. "Data-driven health estimation and lifetime prediction of lithium-ion batteries: A review." In: *Renewable and Sustainable Energy Reviews* 113 (2019). ISSN: 13640321. https://doi.org/10.1016/j.rser.2019.109254.
125. Samveg Saxena, Caroline Le Floch, Jason MacDonald, and Scott Moura. "Quantifying EV battery end-of-life through analysis of travel needs with vehicle powertrain models."

In: *Journal of Power Sources* 282 (6064 2015), pp. 265–276. ISSN: 03787753. https://doi.org/10.1016/j.jpowsour.2015.01.072.

126. Yongzhi Zhang, Rui Xiong, Hongwen He, and Weixiang Shen. "Lithium-Ion Battery Pack State of Charge and State of Energy Estimation Algorithms Using a Hardware-in-the-Loop Validation." In: *IEEE Transactions on Power Electronics* 32 (6 2017), pp. 4421–4431. ISSN: 0885-8993. https://doi.org/10.1109/TPEL.2016.2603229.
127. Bundesministeriums der Justiz. *Verordnung über Verbraucherinformationen zu Kraftstoffverbrauch, CO 2 -Emissionen und Stromverbrauch neuer Personenkraftwagen (Pkw-Energieverbrauchskennzeichnungsverordnung)*. Pkw-EnVKV. Version 2020. 2004. url: https://www.gesetze-im-internet.de/pkw-envkv/__1.html (visited on 01.03.2023).
128. Tadashi Tanaka, S. Ito, M. Muramatsu, Tsutomu Yamada, H. Kamiko, Naoto Kakimoto, and Yoshitaka Inui. "Accurate and versatile simulation of transient voltage profile of lithium-ion secondary battery employing internal equivalent electric circuit." In: *Applied Energy* 143 (3 2015), pp. 200–210. ISSN: 1872-9118. https://doi.org/10.1016/j.apenergy.2015.01.028.
129. United Nations Economic Commission for Europe. *Addendum 22: United Nations Global Technical Regulation (GTR) No. 22. United Nations Global Technical Regulation on In-vehicle Battery Durability for Electrified Vehicles*. 2022. url: https://unece.org/sites/default/files/2022-04/ECE_TRANS_180a22e.pdf (visited on 01.03.2023).
130. United Nations Economic Commission for Europe. *Major auto markets join forces for draft UN legislation on electric vehicle battery durability*. 2021. url: https://unece.org/media/press/362038 (visited on 01.03.2023).
131. United Nations Economic Commission for Europe. *UBE Provides Battery Durability Insight*. 2020. url: https://wiki.unece.org/download/attachments/109351800/EVE-38-06e.pdf (visited on 01.03.2023).
132. Lisa Calearo, Charalampos Ziras, Andreas Thingvad, and Mattia Marinelli. "Agnostic Battery Management System Capacity Estimation for Electric Vehicles." In: *Energies* 15 (24 2022), p. 9656. ISSN: 1996-1073. https://doi.org/10.3390/en15249656.
133. Carlos Pastor-Fernandez, Thomas Bruen, Widanalage Dhammika Widanage, M. A. Gama-Valdez, and James Marco. "A Study of Cell-to-Cell Interactions and Degradation in Parallel Strings. Implications for the Battery Management System." In: *Journal of Power Sources* 329 (0 2016), pp. 574–585. ISSN: 03787753. https://doi.org/10.1016/j.jpowsour.2016.07.121.
134. Eberhard Meissner and Gerolf Richter. "Battery Monitoring and Electrical Energy Management." In: *Journal of Power Sources* 116 (1-2 2003), pp. 79–98. ISSN: 03787753. https://doi.org/10.1016/S0378-7753(02)00713-9.
135. Yujie Wang, Jiaqiang Tian, Zhendong Sun, Li Wang, Ruilong Xu, Mince Li, and Zonghai Chen. "A comprehensive review of battery modeling and state estimation approaches for advanced battery management systems." In: *Renewable and Sustainable Energy Reviews* 131 (2 2020), p. 110015. ISSN: 13640321. https://doi.org/10.1016/j.rser.2020.110015.
136. Mohammad Javad Esfandyari, Vahid Esfahanian, Mohammad Reza Ha'iri-Yazdi, Hassan Nehzati, and Omid Shekoofa. "A new approach to consider the influence of aging state on Lithium-ion battery state of power estimation for hybrid electric vehic-

le." In: *Energy* 176 (2019), pp. 505–520. ISSN: 1873-6785. https://doi.org/10.1016/j.energy.2019.03.176.

137. Larry W. Juang, Phillip J. Kollmeyer, Thomas M. Jahns, and Robert D. Lorenz. "Implementation of online battery state-of-power and state-of-function estimation in electric vehicle applications." In: *IEEE Energy Conversion Congress and Exposition (ECCE)* (Raleigh, NC, USA, 15.09.–20.09.2012). 2012, pp. 1819–1826. https://doi.org/10.1109/ECCE.2012.6342591.
138. Lluc Canals Casals and Beatriz Amante García. "Second-Life Batteries on a Gas Turbine Power Plant to Provide Area Regulation Services." In: *Batteries* 3 (4 2017), p. 10. ISSN: 2313-0105. https://doi.org/10.3390/batteries3010010.
139. Rodolfo Dufo-López and José L. Bernal-Agustín. "Techno-economic analysis of grid-connected battery storage." In: *Energy Conversion and Management* 91 (2015), pp. 394–404. ISSN: 01968904. https://doi.org/10.1016/j.enconman.2014.12.038.
140. Nikolaos Wassiliadis, Matthias Steinsträter, Markus Schreiber, Philipp Rosner, Lorenzo Nicoletti, Florian Schmid, Manuel Ank, Olaf Teichert, Leo Wildfeuer, Jakob Schneider, Alexander Koch, Adrian König, Andreas Glatz, Josef Gandlgruber, Thomas Kröger, Xue Lin, and Markus Lienkamp. "Quantifying the state of the art of electric powertrains in battery electric vehicles. Range, efficiency, and lifetime from component to system level of the Volkswagen ID.3." In: *eTransportation* 12 (1 2022), p. 100167. ISSN: 25901168. https://doi.org/10.1016/j.etran.2022.100167.
141. Peter Lamp. "Requirements for batteries used in electric mobility applications." In: *Lithium-Ion Batteries: Basics and Applications*. Ed. by Reiner Korthauer. Berlin, Heidelberg, Germany: Springer Berlin Heidelberg, 2018, pp. 371–391. ISBN: 978-3-662-53069-6. https://doi.org/10.1007/978-3-662-53071-9_31.
142. Juner Zhu, Ian Mathews, Dongsheng Ren, Wei Li, Daniel A. Cogswell, Bobin Xing, Tobias Sedlatschek, Sai Nithin R. Kantareddy, Mengchao Yi, Tao Gao, Yong Xia, Qing Zhou, Tomasz Wierzbicki, and Martin Z. Bazant. "End-of-life or second-life options for retired electric vehicle batteries." In: *Cell Reports Physical Sciences* 2 (8 2021), p. 100537. ISSN: 26663864. https://doi.org/10.1016/j.xcrp.2021.100537.
143. Egoitz Martinez-Laserna, Elixabet Sarasketa-Zabala, Igor Villarreal Sarria, Daniel-Ioan Stroe, Maciej Jozef Swierczynski, Alexander Warnecke, Jean-Marc Timmermans, Shovon Goutam, Noshin Omar, and Pedro Rodriguez. "Technical Viability of Battery Second Life. A Study From the Ageing Perspective." In: *IEEE Transactions on Industry Applications* 54 (3 2018), pp. 2703–2713. ISSN: 1939-9367. https://doi.org/10.1109/TIA.2018.2801262.
144. Haochen Qin, Xuexin Fan, Yaxiang Fan, Ruitian Wang, and Fengyuan Tian. "Lithium-ion Batteries RUL Prediction Based on Temporal Pattern Attention." In: *Journal of Physics: Conference Series* 2320 (1 2022), p. 012005. ISSN: 1742-6588. https://doi.org/10.1088/1742-6596/2320/1/012005.
145. Adnan Nuhic, Tarik Terzimehic, Thomas Soczka-Guth, Michael Buchholz, and Klaus Dietmayer. "Health diagnosis and remaining useful life prognostics of lithium-ion batteries using data-driven methods." In: *Journal of Power Sources* 239 (3 2013), pp. 680–688. ISSN: 03787753. https://doi.org/10.1016/j.jpowsour.2012.11.146.
146. Danpeng Cheng, Wuxin Sha, Linna Wang, Shun Tang, Aijun Ma, Yongwei Chen, Huawei Wang, Ping Lou, Songfeng Lu, and Yuan-Cheng Cao. "Solid-State Lithium Battery

Cycle Life Prediction Using Machine Learning." In: *Applied Sciences* 11 (10 2021), p. 4671. ISSN: 2076-3417. https://doi.org/10.3390/app11104671.

147. Miguel A. Ortega-Vazquez. "Optimal scheduling of electric vehicle charging and vehicle-to-grid services at household level including battery degradation and price uncertainty." In: *IET Generation, Transmission & Distribution* 8 (6 2014), pp. 1007–1016. ISSN: 1751-8687. https://doi.org/10.1049/iet-gtd.2013.0624.
148. Florian Wankmüller, Prakash R. Thimmapuram, Kevin G. Gallagher, and Audun Botterud. "Impact of battery degradation on energy arbitrage revenue of grid-level energy storage." In: *Journal of Energy Storage* 10 (11 2017), pp. 56–66. ISSN: 2352152X. https://doi.org/10.1016/j.est.2016.12.004.
149. Madeleine Ecker, Jochen B. Gerschler, Jan Vogel, Stefan Kaebitz, Friedrich Hust, Philipp André Dechent, and Dirk Uwe Sauer. "Development of a lifetime prediction model for lithium-ion batteries based on extended accelerated aging test data." In: *Journal of Power Sources* 215 (2012), pp. 248–257. ISSN: 03787753. https://doi.org/10.1016/j.jpowsour.2012.05.012.
150. Samuel Greenbank and David A. Howey. "Automated Feature Extraction and Selection for Data-Driven Models of Rapid Battery Capacity Fade and End of Life." In: *IEEE Transactions on Industrial Informatics* 18 (5 2022), pp. 2965–2973. ISSN: 1551-3203. https://doi.org/10.1109/TII.2021.3106593.
151. Heinz Wenzl, Ian Baring-Gould, Rudi Kaiser, Boryann Liaw, Per Lundsager, Jim Manwell, Alan J. Ruddell, and Vojtech Svoboda. "Life prediction of batteries for selecting the technically most suitable and cost effective battery." In: *Journal of Power Sources* 144 (2 2005), pp. 373–384. ISSN: 03787753. https://doi.org/10.1016/j.jpowsour.2004.11.045.
152. Andrew K.S. Jardine, Daming Lin, and Dragan Banjevic. "A review on machinery diagnostics and prognostics implementing condition-based maintenance." In: *Mechanical Systems and Signal Processing* 20 (7 2006), pp. 1483–1510. ISSN: 08883270. https://doi.org/10.1016/j.ymssp.2005.09.012.
153. Jingliang Zhang and Jay Lee. "A review on prognostics and health monitoring of Li-ion battery." In: *Journal of Power Sources* 196 (15 2011), pp. 6007–6014. ISSN: 03787753. https://doi.org/10.1016/j.jpowsour.2011.03.101.
154. Christoph Birkl. "Diagnosis and Prognosis of Degradation in Lithium-Ion Batteries." PhD Thesis. Oxford, UK, 2017. 191 pp. url: https://ora.ox.ac.uk/objects/uuid:7d8ccb9c-1469-4209-9995-5871fc908b54/files/m0f85ce4df327a1b2604cdd066d4110c9 (visited on 01.03.2023).
155. Jens Vetter, Petr Novák, Markus R. Wagner, Claudia Veit, Kai-Christian Möller, Júrgen Otto M. Besenhard, Martin Winter, Margret Wohlfahrt-Mehrens, C. Vogler, and Abderrezak Hammouche. "Ageing mechanisms in lithium-ion batteries." In: *Journal of Power Sources* 147 (1-2 2005), pp. 269–281. ISSN: 03787753. https://doi.org/10.1016/j.jpowsour.2005.01.006.
156. Brian Bole, Chetan S. Kulkarni, and Matthew Daigle. "Adaptation of an Electrochemistry-based Li-Ion Battery Model to Account for Deterioration Observed Under Randomized Use." In: *Annual Conference of the Prognostics and Health Management Society* (Minneapolis, MN, US, 23.09.–27.09.2021). Vol. 5. 2012. https://doi.org/10.36001/phmconf.2014.v6i1.2490.

157. Nathalie Legrand, Bernard Knosp, Philippe Desprez, Françloise Lapicque, and Stéphane Raël. "Physical characterization of the charging process of a Li-ion battery and prediction of Li plating by electrochemical modelling." In: *Journal of Power Sources* 245 (2014), pp. 208–216. ISSN: 03787753. https://doi.org/10.1016/j.jpowsour.2013.06.130.
158. Thomas Waldmann, Björn-Ingo Hogg, and Margret Wohlfahrt-Mehrens. "Li plating as unwanted side reaction in commercial Li-ion cells—A review." In: *Journal of Power Sources* 384 (2018), pp. 107–124. ISSN: 03787753. https://doi.org/10.1016/j.jpowsour.2018.02.063.
159. Tuan Tran Nguyen. "Big Data Alterungsanalyse von Fahrzeugantriebsbatterien zur Klassifizierung für stationäre Anwendungen. Der Einfluss der Fahrweise auf die Battery 2nd Life Verwendung." PhD Thesis. München, Germany, 2019. url: https://mediatum.ub.tum.de/doc/1488578/1488578.pdf (visited on 01.03.2023).
160. Bramy Pilipili Matadi, Sylvie Geniès, Arnaud Delaille, Thomas Waldmann, Michael Kasper, Margret Wohlfahrt-Mehrens, Frederic Aguesse, Emilie Bekaert, Isabel Jiménez-Gordon, Lise Daniel, Xavier Fleury, Michel Bardet, Jean-Frédéric Martin, and Yann Bultel. "Effects of biphenyl polymerization on lithium deposition in commercial graphite/NMC lithium-ion pouch-cells during calendar aging at high temperature." In: *Journal of The Electrochemical Society* 164 (6 2017), A1089-A1097. ISSN: 0013-4651. https://doi.org/10.1149/2.0631706jes.
161. Joris Jaguemont, Loïc Boulon, and Yves Dubé. "A comprehensive review of lithium-ion batteries used in hybrid and electric vehicles at cold temperatures." In: *Applied Energy* 164 (Dec 2016), pp. 99–114. ISSN: 03062619; 1872-9118. https://doi.org/10.1016/j.apenergy.2015.11.034.
162. Jürgen Garche, ed. *Encyclopedia of electrochemical power sources*. 1st ed. Amsterdam, The Netherlands: Elsevier, 2009. 12 pp. ISBN: 978-0-444-52745-5.
163. Andrea Marongiu, Marco Roscher, and Dirk Uwe Sauer. "Influence of the vehicletogrid strategy on the aging behavior of lithium battery electric vehicles." In: *Applied Energy* 137 (2015), pp. 899–912. ISSN: 03062619; 1872-9118. https://doi.org/10.1016/j.apenergy.2014.06.063.
164. Bernd Epding, Björn Rumberg, Hannes Jahnke, Ina Stradtmann, and Arno Kwade. "Investigation of significant capacity recovery effects due to long rest periods during high current cyclic aging tests in automotive lithium ion cells and their influence on lifetime." In: *Journal of Energy Storage* 22 (2019), pp. 249–256. ISSN: 2352152X. https://doi.org/10.1016/j.est.2019.02.015.
165. Meinert Lewerenz, Philipp André Dechent, and Dirk Uwe Sauer. "Investigation of capacity recovery during rest period at different states-of-charge after cycle life test for prismatic Li(Ni1/3Mn1/3Co1/3)O2-graphite cells." In: *Journal of Energy Storage* 21 (2019), pp. 680–690. ISSN: 2352152X. https://doi.org/10.1016/j.est.2019.01.004.
166. Jörn Wilhelm, Stefan Seidlmayer, Peter Keil, Jörg Schuster, Armin Kriele, Ralph Gilles, and Andreas Jossen. "Cycling capacity recovery effect. A coulombic efficiency and post-mortem study." In: *Journal of Power Sources* 365 (2017), pp. 327–338. ISSN: 03787753. https://doi.org/10.1016/j.jpowsour.2017.08.090.
167. Kevin L. Gering, Sergiy V. Sazhin, David K. Jamison, Christopher J. Michelbacher, Boryann Liaw, Matthieu Dubarry, and Mikael Cugnet. "Investigation of path dependence in commercial lithium-ion cells chosen for plug-in hybrid vehicle duty cycle protocols."

In: *Journal of Power Sources* 196 (7 2011), pp. 3395–3403. ISSN: 03787753. https://doi.org/10.1016/j.jpowsour.2010.05.058.

168. Alexander Karger, Leo Wildfeuer, Deniz Aygül, Arpit Maheshwari, Jan P. Singer, and Andreas Jossen. "Modeling capacity fade of lithium-ion batteries during dynamic cycling considering path dependence." In: *Journal of Energy Storage* 52 (2 2022), p. 104718. ISSN: 2352152X. https://doi.org/10.1016/j.est.2022.104718.
169. Trishna Raj, Andrew A. Wang, Charles W. Monroe, and David A. Howey. "Investigation of Path-Dependent Degradation in Lithium-Ion Batteries." In: *Batteries & Supercaps* 3 (12 2020), pp. 1377–1385. ISSN: 2566-6223. https://doi.org/10.1002/batt.202000160.
170. Laisuo Su, Jianbo Zhang, Jun Huang, Hao Ge, Zhe Li, Fengchao Xie, and Boryann Liaw. "Path dependence of lithium ion cells aging under storage conditions." In: *Journal of Power Sources* 315 (9 2016), pp. 35–46. ISSN: 03787753. https://doi.org/10.1016/j.jpowsour.2016.03.043.
171. Kai Schofer, Florian Laufer, Jochen Stadler, Severin Hahn, Gerd Gaiselmann, Arnulf Latz, and Kai Peter Birke. "Machine Learning-Based Lifetime Prediction of Lithium-Ion Cells." In: *Advanced science (Weinheim, Baden-Wurttemberg, Germany)* (2022), e2200630. https://doi.org/10.1002/advs.202200630.
172. Paula Fermłn-Cueto, Euan McTurk, Michael Allerhand, Encarni Medina-Lopez, Miguel F. Anjos, Joel Sylvester, and Gonçalo dos Reis. "Identification and machine learning prediction of knee-point and knee-onset in capacity degradation curves of lithium-ion cells." In: *Energy and AI* 1 (5 2020), p. 100006. ISSN: 26665468. https://doi.org/10.1016/j.egyai.2020.100006.
173. Peter M. Attia, Alexander Bills, Ferran Brosa Planella, Philipp André Dechent, Gonçalo dos Reis, Matthieu Dubarry, Paul Gasper, Richard Gilchrist, Samuel Greenbank, David A. Howey, Ouyang Liu, Edwin Khoo, Yuliya Preger, Abhishek Soni, Shashank Sripad, Anna G. Stefanopoulou, and Valentin Sulzer. "Review-"Knees" in Lithium-Ion Battery Aging Trajectories." In: *Journal of The Electrochemical Society* 169 (6 2022), p. 060517. ISSN: 0013-4651. https://doi.org/10.1149/1945-7111/ac6d13.
174. Alexander Warnecke. "Degradation Mechanisms in NMC Based Lithium-Ion Batteries." Institute for Power Electronics and Electrical Drives (ISEA). PhD Thesis. Aachen, Germany, 2017. 173 pp. https://doi.org/10.18154/RWTH-2017-09646.
175. Xiao-Guang Yang, Yongjun Leng, Guangsheng Zhang, Shanhai Ge, and Chao-Yang Wang. "Modeling of lithium plating induced aging of lithium-ion batteries: Transition from linear to nonlinear aging." In: *Journal of Power Sources* 360 (2017), pp. 28–40. ISSN: 03787753. https://doi.org/10.1016/j.jpowsour.2017.05.110.
176. Elisa Braco, Idoia San Martin, Alberto Berrueta, Pablo Sanchis, and Alfredo Ursua. "Experimental assessment of cycling ageing of lithium-ion second-life batteries from electric vehicles." In: *Journal of Energy Storage* 32 (2020). ISSN: 2352152X. https://doi.org/10.1016/j.est.2020.101695.
177. Mattin Lucu, Egoitz Martinez-Laserna, Iñigo Gandiaga, Kailong Liu, Haritza Camblong, Widanalage Dhammika Widanage, and James Marco. "Data-driven nonparametric Li-ion battery ageing model aiming at learning from real operation data—Part A. Storage operation." In: *Journal of Energy Storage* 30 (2020), p. 101409. ISSN: 2352152X. https://doi.org/10.1016/j.est.2020.101409.
178. Mathias Petzl, Michael Kasper, and Michael A. Danzer. "Lithium plating in a commercial lithium-ion battery—A low-temperature aging study." In: *Journal of Power Sources*

275 (2015), pp. 799–807. ISSN: 03787753. https://doi.org/10.1016/j.jpowsour.2014.11.065.

179. Hamed Sadegh Kouhestani, Xiaoping Yi, Guoqing Qi, Xunliang Liu, Ruimin Wang, Yang Gao, Xiao Yu, and Lin Liu. "Prognosis and Health Management (PHM) of Solid-State Batteries. Perspectives, Challenges, and Opportunities." In: *Energies* 15 (18 2022), p. 6599. ISSN: 1996-1073. https://doi.org/10.3390/en15186599.
180. Teo Lombardo, Marc Duquesnoy, Hassna El-Bouysidy, Fabian Årén, Alfonso Gallo-Bueno, Peter Bjørn Jørgensen, Arghya Bhowmik, Arnaud Demortière, Elixabete Ayerbe, Francisco Alcaide, Marine Reynaud, Javier Carrasco, Alexis Grimaud, Chao Zhang, Tejs Vegge, Patrik Johansson, and Alejandro A. Franco. "Artificial Intelligence Applied to Battery Research. Hype or Reality?" In: *Chemical reviews* 122 (12 2022), pp. 10899–10969. ISSN: 0009-2665. https://doi.org/10.1021/acs.chemrev.1c00108.
181. International Organization for Standardization. ISO/IEC 62620:2014. 2014. *Secondary cells and batteries containing alkaline or other non-acid electrolytes – Secondary lithium cells and batteries for use in industrial applications.*
182. Antti Väyrynen and Justin Salminen. "Lithium ion battery production." In: *The Journal of Chemical Thermodynamics* 46 (Spring 2012), pp. 80–85. ISSN: 00219614. https://doi.org/10.1016/j.jct.2011.09.005.
183. Arno Kwade, Wolfgang Haselrieder, Ruben Leithoff, Armin Modlinger, Franz Dietrich, and Klaus Droeder. "Current status and challenges for automotive battery production technologies." In: *Nature Energy* 3 (4 2018), pp. 290–300. ISSN: 2058-7546. https://doi.org/10.1038/s41560-018-0130-3.
184. Achim Kampker. *Elektromobilproduktion*. Berlin, Heidelberg, Germany: Springer Berlin Heidelberg, 2014. 290 pp. https://doi.org/10.1007/978-3-642-42022-1.
185. Hendrik Löbberding, Saskia Wessel, Christian Offermanns, Mario Kehrer, Johannes Rother, Heiner Hans Heimes, and Achim Kampker. "From Cell to Battery System in BEVs. Analysis of System Packing Efficiency and Cell Types." In: *World Electric Vehicle Journal* 11 (4 2020), p. 77. ISSN: 2032-6653. https://doi.org/10.3390/wevj11040077.
186. Andreas Hauser and Robert M. Kuhn. "High-voltage battery management systems (BMS) for electric vehicles." In: *Advances in battery technologies for electric vehicles*. Ed. by Bruno Scrosati, Jürgen Garche, and Werner Tillmetz. 1st ed. Woodhead Publishing series in energy 80. Cambridge, UK:Woodhead Publishing, 2015, pp. 265–282. ISBN: 9781782423775. https://doi.org/10.1016/B978-1-78242-377-5.00011-X.
187. Xiaoxi Zhang, Yue Xiong, Yongjun Pan, Haifeng Du, and Binghe Liu. "Crushing stress and vibration fatigue-life optimization of a battery-pack system." In: *Struct. Multidisc. Optim. (Structural and Multidisciplinary Optimization)* 66 (3 2023), p. 30. https://doi.org/10.1007/s00158-023-03510-2.
188. Andreas Jossen, V. Späth, Harry Döring, and Jürgen Garche. "Reliable battery operation—a challenge for the battery management system." In: *Journal of Power Sources* 84 (2 1999), pp. 283–286. ISSN: 03787753. https://doi.org/10.1016/S0378-7753(99)00329-8.
189. Michael Baumann, Leo Wildfeuer, Stephan Rohr, and Markus Lienkamp. "Parameter variations within Li-Ion battery packs – Theoretical investigations and experimental quantification." In: *Journal of Energy Storage* 18 (23 2018), pp. 295–307. ISSN: 2352152X. https://doi.org/10.1016/j.est.2018.04.031.

190. Fei Feng, Xiaosong Hu, Lin Hu, Fengling Hu, Yang Li, and Lei Zhang. "Propagation mechanisms and diagnosis of parameter inconsistency within Li-Ion battery packs." In: *Renewable and Sustainable Energy Reviews* 112 (2019), pp. 102–113. ISSN: 13640321. https://doi.org/10.1016/j.rser.2019.05.042.
191. Andrea Carolina Cordoba-Arenas, Simona Onori, and Giorgio Rizzoni. "A control-oriented lithium-ion battery pack model for plug-in hybrid electric vehicle cycle-life studies and system design with consideration of health management." In: *Journal of Power Sources* 279 (2015), pp. 791–808. ISSN: 03787753. https://doi.org/10.1016/j.jpowsour.2014.12.048.
192. Chunxiao Luan, Chen Ma, Chunyu Wang, Long Chang, Linjing Xiao, Zhihao Yu, and Hongyu Li. "Influence of the connection topology on the performance of lithium-ion battery pack under cell-to-cell parameters variations." In: *Journal of Energy Storage* 41 (2021), p. 102896. ISSN: 2352152X. https://doi.org/10.1016/j.est.2021.102896.
193. Gregory L. Plett and Martin J. Klein. "Simulating Battery Packs Comprising Parallel Cell Modules and Series Cell Modules. EVS24 International Battery, Hybrid and Fuel Cell Electric Vehicle Symposium." In: *24th International Battery, Hybrid and Fuel Cell Electric Vehicle Symposium & Exhibition (EVS 24)* (Stavanger, Norway, 13.05.–16.05.2009). 2009, pp. 1–17. url: https://www.evs24.org/php/download.php?f=papers/EVS-24-530150.pdf.
194. Olof Juhlin. "Modeling of Battery Degradation in Electrified Vehicles." Department of Electrical Engineering, Vehicular Systems. Master Thesis. Linköping, Sweden, 2016. url: http://urn.kb.se/resolve?urn=urn:nbn:se:liu:diva-134114 (visited on 01.03.2023).
195. Yin Hua, Andrea Carolina Cordoba-Arenas, Nicholas Warner, and Giorgio Rizzoni. "A multi time-scale state-of-charge and state-of-health estimation framework using non-linear predictive filter for lithium-ion battery pack with passive balance control." In: *Journal of Power Sources* 280 (1 2015), pp. 293–312. ISSN: 03787753. https://doi.org/10.1016/j.jpowsour.2015.01.112.
196. Dirk Uwe Sauer. *Einfluss der Batteriestreuung und Lebensdauer auf das Batteriesystemdesign*. 3. Elektromobilitätsproduktionstag, Aachen, 28.10.2015. 2015. https://doi.org/10.13140/RG.2.1.1417.1283. (visited on 01.03.2023).
197. Shuen-Lin Jeng, Cher Ming Tan, and Ping-Chia Chen. "Statistical distribution of Lithium-ion batteries useful life and its application for battery pack reliability." In: *Journal of Energy Storage* 51 (15 2022), p. 104399. ISSN: 2352152X. https://doi.org/10.1016/j.est.2022.104399.
198. Isidor Buchmann. *BU-302: Series and Parallel Battery Configurations*. 2021. url: https://batteryuniversity.com/article/bu-302-series-and-parallel-battery-configurations (visited on 01.03.2023).
199. Chin-Yao Chang, Punit Tulpule, Giorgio Rizzoni, Wei Zhang, and Xinyu Du. "A probabilistic approach for prognosis of battery pack aging." In: *Journal of Power Sources* 347 (2 2017), pp. 57–68. ISSN: 03787753. https://doi.org/10.1016/j.jpowsour.2017.01.130.
200. Sebastian Paul, Christian Diegelmann, Herbert Kabza, and Werner Tillmetz. "Analysis of ageing inhomogeneities in lithium-ion battery systems." In: *Journal of Power Sources* 239 (2013), pp. 642–650. ISSN: 03787753. https://doi.org/10.1016/j.jpowsour.2013.01.068.

201. Matthieu Dubarry, George Baure, Carlos Pastor-Fernandez, Tung Fai Yu, Widanalage Dhammika Widanage, and James Marco. "Battery energy storage system modeling: A combined comprehensive approach." In: *Journal of Energy Storage* 21 (2019), pp. 172–185. ISSN: 2352152X. https://doi.org/10.1016/j.est.2018.11.012.
202. Thomas Bruen and James Marco. "Modelling and experimental evaluation of parallel connected lithium ion cells for an electric vehicle battery system." In: *Journal of Power Sources* 310 (2016), pp. 91–101. ISSN: 03787753. https://doi.org/10.1016/j.jpowsour.2016.01.001.
203. Dae Hyun Jung, Dong Min Kim, Jonghoo Park, Sang-il Kim, and TaeWan Kim. "Cycle-life prediction model of lithium iron phosphate-based lithium-ion battery module." In: *International Journal of Energy Research* 45 (11 2021), pp. 16489–16496. ISSN: 1099-114X. https://doi.org/10.1002/er.6895.
204. Tanvir R. Tanim, Matthew G. Shirk, Randy L. Bewley, Eric J. Dufek, and Boryann Liaw. "Fast charge implications. Pack and cell analysis and comparison." In: *Journal of Power Sources* 381 (4 2018), pp. 56–65. ISSN: 03787753. https://doi.org/10.1016/j.jpowsour.2018.01.091.
205. Lisa Calearo, Andreas Thingvad, Charalampos Ziras, and Mattia Marinelli. "A methodology to model and validate electro-thermal-aging dynamics of electric vehicle battery packs." In: *Journal of Energy Storage* 55 (12 2022), p. 105538. ISSN: 2352152X. https://doi.org/10.1016/j.est.2022.105538.
206. Matthieu Dubarry and David A.C. Beck. "Perspective on Mechanistic Modeling of Li-Ion Batteries." In: *ACCOUNTS OF MATERIALS RESEARCH* 3 (8 2022), pp. 843–853. ISSN: 2643-6728. https://doi.org/10.1021/accountsmr.2c00082.
207. Max Peter Naylor Marlow. "Understanding the effects of heterogeneities on battery pack lifetime." Design Engineering. PhD Thesis. London, UK, 2021. 248 pp. url: https://doi.org/10.25560/92251.
208. Anna Tomaszewska, Zhengyu Chu, Xuning Feng, Simon O'Kane, Xinhua Liu, Jingyi Chen, Chenzhen Ji, Elizabeth Endler, Ruihe Li, Lishuo Liu, Yalun Li, Siqi Zheng, Sebastian Vetterlein, Ming Gao, Jiuyu Du, Michael Parkes, Minggao Ouyang, Monica Marinescu, Gregory J. Offer, and Billy Wu. "Lithium-ion battery fast charging. A review." In: *eTransportation* 1 (15 2019), p. 100011. ISSN: 25901168. https://doi.org/10.1016/j.etran.2019.100011.
209. AUDI AG. *Charging technologies.* 2018. url: https://www.audi-mediacenter.com/en/technology-lexicon-7180/charging-technologies-10999 (visited on 01.03.2023).
210. Paul Gasper and Kandler Smith. *Predictive Battery Lifetime Modeling at the National Renewable Energy Laboratory.* 2021. url: https://www.osti.gov/biblio/1838001 (visited on 01.03.2023).
211. Dirk Uwe Sauer. *Time-series cyclic aging data on 48 commercial NMC/ graphite Sanyo/ Panasonic UR18650E cylindrical cells.* 2021. https://doi.org/10.18154/RWTH-2021-04545. (visited on 01.03.2023).
212. Valentin Sulzer, Peyman Mohtat, Antti Aitio, Suhak Lee, Yen T. Yeh, Frank Steinbacher, Muhammad Umer Khan, Jang Woo Lee, Jason B. Siegel, Anna G. Stefanopoulou, and David A. Howey. "The challenge and opportunity of battery lifetime prediction from field data." In: *Joule* 5 (8 2021), pp. 1934–1955. ISSN: 2542-4351. https://doi.org/10.1016/j.joule.2021.06.005.

213. Fangfang Yang, Xiangbao Song, Guangzhong Dong, and Kwok-Leung Tsui. "A coulombic efficiency-based model for prognostics and health estimation of lithium-ion batteries." In: *Energy* 171 (2019), pp. 1173–1182. ISSN: 1873-6785. https://doi.org/10.1016/j.energy.2019.01.083.
214. Sébastien Grolleau, Issam Baghdadi, Philippe Gyan, Mohamed Ben-Marzouk, and François Duclaud. "Capacity Fade of Lithium-Ion Batteries upon Mixed Calendar/Cycling Aging Protocol." In: *World Electric Vehicle Journal* 8 (2 2016), pp. 339–349. ISSN: 2032-6653. https://doi.org/10.3390/wevj8020339.
215. Rasheed Ibraheem, Calum Strange, and Gonçalo dos Reis. "Capacity and Internal Resistance of lithium-ion batteries. Full degradation curve prediction from Voltage response at constant Current at discharge." In: *Journal of Power Sources* 556 (5 2023), p. 232477. ISSN: 03787753. https://doi.org/10.1016/j.jpowsour.2022.232477.
216. Erik Figenbaum. "Battery Electric Vehicle Fast Charging-Evidence from the Norwegian Market." In: *World Electric Vehicle Journal* 11 (2 2020), p. 38. ISSN: 2032-6653. https://doi.org/10.3390/wevj11020038.
217. Quentin Badey, Guillaume Cherouvrier, Yvan Reynier, Jean-Marc Duffault, and Sylvain Franger. "Ageing forecast of lithium-ion batteries for electric and hybrid vehicles." In: *Current Topics in Electrochemistry* (16 2011), pp. 65–79. ISSN: 0972-4443.
218. Jingxi Yang, Matthew Beatty, Dani Strickland, Mina Abedi-Varnosfaderani, and Joe Warren. "Second Life Battery Life Cycle Estimation." In: (2022). https://doi.org/10.36227/techrxiv.21610080.
219. Malte F. Jung, David Sirkin, Turgut M. Gür, and Martin Steinert. "Displayed Uncertainty Improves Driving Experience and Behavior." In: *CHI '15: CHI Conference on Human Factors in Computing Systems* (Seoul, Republic of Korea, 18.04.–23.04.2015). Ed. by Bo Begole, Jinwoo Kim, Kori Inkpen, and Woontack Woo. 2015, pp. 2201–2210. https://doi.org/10.1145/2702123.2702479.
220. Fraunhofer-Institut für System- und Innovationsforschung. *Technologie-Roadmap Stationäre Energiespeicher 2030*. 2015. url: https://www.isi.fraunhofer.de/content/dam/isi/dokumente/cct/lib/TRM-SES.pdf (visited on 01.03.2023).
221. Benedikt Tepe, Jan Figgener, Stefan Englberger, Dirk Uwe Sauer, Andreas Jossen, and Holger Hesse. "Optimal pool composition of commercial electric vehicles in V2G fleet operation of various electricity markets." In: *Applied Energy* 308 (1 2022), p. 118351. ISSN: 03062619; 1872-9118. https://doi.org/10.1016/j.apenergy.2021.118351.
222. Milad Ghorbanzadeh, Majid Astaneh, and Farzin Golzar. "Long-term degradation based analysis for lithium-ion batteries in off-grid wind-battery renewable energysystems." In: *Energy* 166 (2019), pp. 1194–1206. ISSN: 1873-6785. https://doi.org/10.1016/j.energy.2018.10.120.
223. Matthieu Dubarry, Moe Tun, George Baure, Marc Matsuura, and Richard E. Rocheleau. "Battery Durability and Reliability under Electric Utility Grid Operations: Analysis of On-Site Reference Tests." In: *Electronics* 10 (13 2021). ISSN: 2079-9292. https://doi.org/10.3390/electronics10131593.
224. Guzay Pasaoglu Kilanc, Davide Fiorello, Aangelo Martino, Gabriella Scarcella, Andrea Alemanno, Alyona Zubaryeva, and Christian Thiel. *Driving and parking patterns of European car drivers. A mobility survey.* Vol. 25627. EUR. Scientific and technical research series. Luxembourg: Publications Office of the European Union, 2012. ISBN: 978-92-79-27739-9. https://doi.org/10.2790/7028.

225. Miloš N. Mladenović. "Mobility as a Service." In: *International Encyclopedia of Transportation*. Ed. by Roger Vickerman. 1st ed. San Diego, CA, US: Elsevier, 2021, pp. 12–18. ISBN: 9780081026724. https://doi.org/10.1016/B978-0-08-102671-7.10607-4.
226. Je-Feng Li, Chi-Hao Lin, and Kuo-Ching Chen. "Cycle Life Prediction of Aged Lithium-Ion Batteries from the Fading Trajectory of a Four-Parameter Model." In: *Journal of The Electrochemical Society* 165 (16 2018), A3634-A3641. ISSN: 0013-4651. https://doi.org/10.1149/2.0211816jes.
227. George Baure and Matthieu Dubarry. "Synthetic vs. Real Driving Cycles. A Comparison of Electric Vehicle Battery Degradation." In: *Batteries* 5 (2 2019), p. 42. ISSN: 2313-0105. https://doi.org/10.3390/batteries5020042.
228. Daniel-Ioan Stroe, Maciej Jozef Swierczynski, Søren Knudesn Kær, Egoitz Martinez-Laserna, and Elixabet Sarasketa-Zabala. "Accelerated aging of Lithium-ion batteries based on electric vehicle mission profile." In: *ECCE 2017. IEEE Energy Conversion Congress & Expo* (Cincinnati, OH, US, 01.10.–05.10.2017). Piscataway, NJ, US: IEEE, 2017. ISBN: 978-1-5090-2998-3. https://doi.org/10.1109/ECCE.2017.8096937.
229. Md Sazzad Hosen, Joris Jaguemont, Joeri van Mierlo, and Maitane Berecibar. "Battery lifetime prediction and performance assessment of different modeling approaches." In: *iScience* 24 (2 2021). https://doi.org/10.1016/j.isci.2021.102060.
230. Salvatore Micari, Salvatore Foti, Antonio Testa, S. de Caro, Francesco Sergi, Laura Andaloro, Davide Aloisio, and Giuseppe Napoli. "Reliability assessment and lifetime prediction of Li-ion batteries for electric vehicles." In: *Electrical Engineering* 104 (1 2022), pp. 165–177. ISSN: 0948-7921. https://doi.org/10.1007/s00202-021-01288-4.
231. Gabriele Pozzato, Anirudh Allam, and Simona Onori. "Lithium-ion battery aging dataset based on electric vehicle real-driving profiles." In: *Data in brief* 41 (2022), p. 107995. https://doi.org/10.1016/j.dib.2022.107995.
232. Julius Schmitt, Mathias Rehm, Alexander Karger, and Andreas Jossen. *Aging data, charging curves and battery electric vehicle application profiles of a nickel-rich NMC / silicon-graphite high-energy lithium-ion cell.* 2023. https://doi.org/10.14459/2022mp1690455.
233. Jochen Stadler, Carsten Krupp, Madeleine Ecker, Jochen Bandlow, Bernd Spier, and Arnulf Latz. "Investigation and modeling of cyclic aging using a design of experiment with automotive grade lithium-ion cells." In: *Journal of Power Sources* 521 (2022), p. 230952. ISSN: 03787753. https://doi.org/10.1016/j.jpowsour.2021.230952.
234. Lars von Kolzenberg, Jochen Stadler, Johannes Fath, Madeleine Ecker, Birger Horstmann, and Arnulf Latz. "A four parameter model for the solid-electrolyte interphase to predict battery aging during operation." In: *Journal of Power Sources* 539 (2022). ISSN: 03787753. https://doi.org/10.1016/j.jpowsour.2022.231560.
235. Anthony Barré, Frédéric Suard, Mathias Gérard, Maxime Montaru, and Delphine Riu. "Statistical analysis for understanding and predicting battery degradations in real-life electric vehicle use." In: *Journal of Power Sources* 245 (9 2014), pp. 846–856. ISSN: 03787753. https://doi.org/10.1016/j.jpowsour.2013.07.052.
236. Penelope K. Jones, Ulrich Stimming, and Alpha A. Lee. "Impedance-based forecasting of lithium-ion battery performance amid uneven usage." In: *Nature Communications* 13 (1 2022). ISSN: 2041-1723. https://doi.org/10.1038/s41467-022-32422-w.
237. Matti Huotari, Shashank Arora, Avleen Malhi, and Kary Främling. "A Dynamic Battery State-of-Health Forecasting Model for Electric Trucks. Li-Ion Batteries Case-Study."

In: *Volume 8: Energy*. ASME 2020 International Mechanical Engineering Congress and Exposition (Virtual, Online, 16.11.–19.11.2020). American Society of Mechanical Engineers, 2020. ISBN: 978-0-7918-8456-0. https://doi.org/10.1115/IMECE2020-23949.

238. Miguel Simão, Rune Prytz, and Sławomir Nowaczyk. "Long-term Evaluation of the State-of-Health of Traction Lithium-ion Batteries in Operational Buses." In: *International Journal of Prognostics and Health Management* 13 (1 2022). ISSN: 2153-2648. https://doi.org/10.36001/?phm.2022.v13i1.3115.

239. Ivana Semanjski and Sidharta Gautama. "Forecasting the State of Health of Electric Vehicle Batteries to Evaluate the Viability of Car Sharing Practices." In: *Energies* 9 (12 2016), p. 1025. ISSN: 1996-1073. https://doi.org/10.3390/en9121025.

240. Fang Yao, Defang Meng, Youxi Wu, Yakun Wan, and Fei Ding. "Online health estimation strategy with transfer learning for operating lithium-ion batteries." In: *Journal of Power Electronics* 239 (5 2023), p. 1526. ISSN: 1598-2092. https://doi.org/10.1007/s43236-023-00605-3.

241. United Nations Economic Commission for Europe. *Addendum 15: Global technical regulation (GTR) No. 15. Global technical regulation on Worldwide harmonized Light vehicles Test Procedure (WLTP).* 2019. url: https://unece.org/fileadmin/DAM/trans/main/wp29/wp29wgs/wp29gen/wp29registry/ECE-TRANS-180a15am5e.pdf (visited on 01.03.2023).

242. United States Environmental Protection Agency. *Dynamometer Drive Schedules.* 2021. url: https://www.epa.gov/vehicle-and-fuel-emissions-testing/dynamometerdrive-schedules#UDDS (visited on 01.03.2023).

243. Fabian Kai-Dietrich Noering. *Unsupervised Pattern Discovery in Automotive Time Series.* 1st ed. AutoUni—Schriftenreihe. Wiesbaden, Germany: Springer Fachmedien Wiesbaden, 2022. ISBN: 978-3-658-36335-2. https://doi.org/10.1007/978-3-658-36336-9.

244. Georg Steinbuß, Benedikt Rzepka, Simon Bischof, Thomas Blank, and Klemens Böhm. "FOBSS: Monitoring Data from a Modular Battery System." In: *e-Energy '19: Proceedings of the Tenth ACM International Conference on Future Energy Systems* (Phoenix, AZ, US, 25.06.–28.06.2019). New York, NY, US: Association for Computing Machinery Inc. (ACM), 2019, pp. 456–459. ISBN: 978-1-4503-6671-7. https://doi.org/10.1145/3307772.3331020.

245. Klaas de Craemer and Khiem Trad. *Cyclic ageing with driving profile of a lithium ion battery module.* Ed. by 4TU.ResearchData. 2021. https://doi.org/10.4121/14096567.v1. (visited on 01.03.2023).

246. Warren S. McCulloch and Walter H. Pitts. "A logical calculus of the ideas immanent in nervous activity." In: *The bulletin of mathematical biophysics* 5 (4 1943), pp. 115–133.

247. Ian Goodfellow, Yoshua Bengio, and Aaron Courville. *Deep learning.* Vol. 2017: 1. Adaptive computation and machine learning. Cambridge, MS, US: The MIT Press, 2017. 775 pp. ISBN: 0262035618.

248. Kurt Hornik, Maxwell Stinchcombe, and Halbert White. "Multilayer feedforward networks are universal approximators." In: *Neural Networks* 2 (5 1989), pp. 359–366. ISSN: 08936080. https://doi.org/10.1016/0893-6080(89)90020-8.

249. Simon S. Haykin. *Neural networks and learning machines.* 3rd ed. New York, NY, US: Pearson, 2009. 906 pp. ISBN: 978-0131471399.

250. Andrew L. Maas, Awni Y. Hannun, and Andrew Y. Ng. "Rectifier nonlinearities improve neural betwork acoustic models." ICML Workshop on Deep Learning for Audio, Speech, and Language Processing. In: *Proceedings of the 30th International Conference on Machine Learning*. International Conference on Machine Learning (ICML) (Atlanta, GA, US, 17.06.–19.06.2013). Ed. by Sanjoy Dasgupta and David McAllester. Proceedings of Machine Learning Research 28. PMLR, 2013. url: https://awnihannun.com/papers/relu_hybrid_icml2013_final.pdf.
251. Yann LeCun, Yoshua Bengio, and Geoffrey E. Hinton. "Deep learning." In: *Nature* 521 (7553 2015), pp. 436–444. https://doi.org/10.1038/nature14539.
252. Shuai Li, Wanqing Li, Chris Cook, Ce Zhu, and Yanbo Gao. "Independently Recurrent Neural Network (IndRNN). Building A Longer and Deeper RNN." In: *IEEE Conference on Computer Vision and Pattern Recognition (CVPR)*. 2018 IEEE/CVF Conference on Computer Vision and Pattern Recognition (CVPR) (Salt Lake City, UT, US, 18.06.–23.06.2018). IEEE Conference on Computer Vision and Pattern Recognition. IEEE, 2018, pp. 5457–5466. ISBN: 978-1-5386-6420-9. https://doi.org/10.1109/CVPR.2018.00572.
253. Razvan Pascanu, Tomas Mikolov, and Yoshua Bengio. "On the difficulty of training Recurrent Neural Networks." In: *Proceedings of the 30th International Conference on Machine Learning*. International Conference on Machine Learning (ICML) (Atlanta, GA, US, 17.06.–19.06.2013). Ed. by Sanjoy Dasgupta and David McAllester. Vol. 28. Proceedings of Machine Learning Research 28. PMLR, 2013, pp. 1310–1318. https://doi.org/10.48550/arXiv.1211.5063.
254. Kyunghyun Cho, Bart van Merrienboer, Caglar Gulcehre, Dzmitry Bahdanau, Fethi Bougares, Holger Schwenk, and Yoshua Bengio. "Learning Phrase Representations using RNN Encoder-Decoder for Statistical Machine Translation." In: *Proceedings of the 2014 Conference on Empirical Methods in Natural Language Processing (EMNLP)*. Proceedings of the 2014 Conference on Empirical Methods in Natural Language Processing (EMNLP) (Doha, Qatar, 01.10.2014). Ed. by Alessandro Moschitti, Bo Pang, and Walter Daeleman. Stroudsburg, PA, US: Association for Computational Linguistics, 2014, pp. 1724–1734. https://doi.org/10.3115/v1/d14-1179.
255. Sepp Hochreiter and Jürgen Schmidhuber. "Long Short-Term Memory." In: *Neural Computation* 9 (8 1997), pp. 1735–1780. ISSN: 0899-7667. https://doi.org/10.1162/neco.1997.9.8.1735.
256. Christopher M. Bishop. *Pattern recognition and machine learning*. 8th ed. Information science and statistics. New York, NY, US: Springer, 2009. 738 pp. ISBN: 978-0-387-31073-2.
257. Yoshua Bengio. "Practical Recommendations for Gradient-Based Training of Deep Architectures." In: *Neural networks. Tricks of the trade*. Ed. by Grégoire Montavon, Geneviéve B. Orr, and Klaus-Robert Müller. 2nd ed. Lecture Notes in Computer Science 7700. Berlin, Germany: Springer, 2012, pp. 437–478. ISBN: 978-3-642-35288-1. https://doi.org/10.1007/978-3-642-35289-8_26.
258. Dominic Masters and Carlo Luschi. *Revisiting small batch training for deep neural Networks*. 2018. https://doi.org/10.48550/arXiv.1804.07612.
259. Sebastian Ruder. *An overview of gradient descent optimization algorithms*. 2017. https://doi.org/10.48550/arXiv.1609.04747.

260. Sergey Ioffe and Christian Szegedy. "Batch Normalization. Accelerating Deep Network Training by Reducing Internal Covariate Shift." In: *32nd International Conference on Machine Learning (ICML 2015)* (Lile, France, 06.07.–11.07.2015). Ed. by Francis Bach and David Blei. Vol. 37. Proceedings of Machine Learning Research. International Conference on Machine Learning. Red Hook, NY, US: Curran Associates Inc., 2016, pp. 448–456. ISBN: 9781510810587. url: https://proceedings.mlr.press/v37/ioffe15.html.
261. Ismail Bin Mohamad and Dauda Usman. "Standardization and its effects on k-means clustering algorithm." In: *Research Journal of Applied Sciences, Engineering and Technology* 6 (17 2013), pp. 3299–3303. ISSN: 20407459. https://doi.org/10.19026/rjaset.6.3638.
262. Joel Grus. *Data science from scratch. First principles with Python.* Revision history for the first edition, second release. Beijing et al.: O'Reilly, 2017. XVI, 311 Seiten. ISBN: 149190142X.
263. Marc Claesen and Bart De Moor. "Hyperparameter search in machine learning." In: *XI Metaheuristics International Conference in Agadir, Morocco (MIC)* (2015). https://doi.org/10.48550/arXiv.1502.02127.
264. Jasper Snoek, Hugo Larochelle, and Ryan P. Adams. *Practical bayesian optimization of machine learning algorithms.* 2012. https://doi.org/10.48550/arXiv.1206.2944.
265. Bobak Shahriari, Kevin Swersky, Ziyu Wang, Ryan P. Adams, and Nando de Freitas. "Taking the human out of the loop: A review of bayesian optimization." In: *Proceedings of the IEEE* 104 (1 2016), pp. 148–175. ISSN: 0018-9219. https://doi.org/10.1109/JPROC.2015.2494218.
266. James S. Bergstra and Yoshua Bengio. "Random search for hyper-parameter optimization." In: *Journal of Machine Learning Research* 13 (2 2012), pp. 281–305. ISSN: 1532-4435.
267. James S. Bergstra, Daniel Yamins, and David Cox. "Making a science of model search: hyperparameter optimization in hundreds of dimensions for vision architecture." In: *Proceedings of the 30th International Conference on Machine Learning.* International Conference on Machine Learning (ICML) (Atlanta, GA, US, 17.06.–19.06.2013). Ed. by Sanjoy Dasgupta and David McAllester. Vol. 28. Proceedings of Machine Learning Research 28. PMLR, 2013, pp. 115–123. url: https://proceedings.mlr.press/v28/bergstra13.html.
268. James S. Bergstra, Rémi Bardenet, Yoshua Bengio, and Balázs Kégl. "Algorithms for Hyper-Parameter Optimization." In: *25th Annual Conference on Neural Information Processing Systems* 2011 (Granada, Spain, 12.12.–15.12.2011). Ed. by J. Shawe-Taylor, R. Zemel, P. Bartlett, F. Pereira, and Kilian Q.Weinberger. 24. Curran Associates Inc., 2011, pp. 2546–2554. ISBN: 978-1-61839-599-3. url: https://proceedings.neurips.cc/paper/2011/file/86e8f7ab32cfd12577bc2619bc635690-Paper.pdf.
269. Frank Hutter, Holger H. Hoos, and Kevin Leyton-Brown. "Sequential Model-Based Optimization for General Algorithm Configuration." In: *Learning and intelligent optimization. 5th international conference*, LION 5, selected papers (Rome, Italy, 17.01.–21.01.2011). Ed. by Carlos A. Coello Coello. Lecture Notes in Computer Science 6683. International Conference on Learning and Intelligent Optimization. Berlin, Germany: Springer, 2011, pp. 507–523. ISBN: 9783642255656. https://doi.org/10.1007/978-3-642-25566-3_40.

270. Rob J. Hyndman and Anne B. Koehler. "Another look at measures of forecast accuracy." In: *International Journal of Forecasting* 22 (4 2006), pp. 679–688. ISSN: 0169-2070. https://doi.org/10.1016/j.ijforecast.2006.03.001.
271. Claude Sammut and Geoffrey I. Webb. *Encyclopedia of Machine Learning and Data Mining*. 2nd ed. Boston, MA, US: Springer US, 2017. 1341 pp. https://doi.org/10.1007/978-1-4899-7687-1.
272. Klaus Backhaus, Bernd Erichson, Wulff Plinke, and Rolf Weiber. *Multivariate Analysemethoden*. Berlin, Heidelberg, Germany: Springer Berlin Heidelberg, 2018. ISBN: 978-3-662-56654-1. https://doi.org/10.1007/978-3-662-56655-8.
273. Sewall Wright. "Correlation and causation." In: *Journal of Agricultural Research* 20 (1921), pp. 557–585. url: https://handle.nal.usda.gov/10113/IND43966364.
274. Davide Chicco, Matthijs J. Warrens, and Giuseppe Jurman. "The coefficient of determination R-squared is more informative than SMAPE, MAE, MAPE, MSE and RMSE in regression analysis evaluation." In: *PeerJ. Computer science* 7 (2021), e623. https://doi.org/10.7717/peerj-cs.623.
275. Sinno Jialin Pan and Qiang Yang. "A Survey on Transfer Learning." In: *IEEE Transactions on Knowledge and Data Engineering* 22 (10 2010), pp. 1345–1359. ISSN: 1041-4347. https://doi.org/10.1109/TKDE.2009.191.
276. Lisa Torrey and Jude Shavlik. "Transfer Learning." In: *Handbook of Research on Machine Learning Applications and Trends. Algorithms, methods, and techniques*. Ed. by Emilio Soria Olivas, José David Martín Guerrero, Marcelino Martinez-Sober, Jose Rafael Magdalena-Benedito, and Antonio José Serrano López. 1st ed. Hershey, PA, US: IGI Global, 2010, pp. 242–264. ISBN: 1605667668.
277. Nicola Bertoldi, Mauro Cettolo, Marcello Federico, and Christian Buck. "Evaluating the Learning Curve of Domain Adaptive Statistical Machine Translation Systems." In: *Proceedings of the 7th Workshop on Statistical Machine Translation Montréal, Canada, June 7-8, 2012* (2012), pp. 433–441.
278. Tom Viering and Marco Loog. "The Shape of Learning Curves. A Review." In: *IEEE Transactions on Pattern Analysis and Machine Intelligence* PP (2022). ISSN: 1939-3539. https://doi.org/10.1109/TPAMI.2022.3220744.
279. Jason Allan Palmer. *Relative Convexity*. 2003. url: https://sccn.ucsd.edu/~jason/relcon_new.pdf (visited on 01.03.2023).
280. Karen Simonyan and Andrew Zisserman. *Very Deep Convolutional Networks for Large-Scale Image Recognition*. 2014. https://doi.org/10.48550/arXiv.1409.1556.
281. Jason Yosinski, Jeff Clune, Yoshua Bengio, and Hod Lipson. "How transferable are features in deep neural networks?" In: *27th International Conference on Neural Information Processing Systems (NIPS)* (Montreal, Canada, 08.12.–13.12.2014). 2014, pp. 3320–3328. https://doi.org/10.48550/arXiv.1411.1792.
282. Ali Sharif Razavian, Hossein Azizpour, Josephine Sullivan, and Stefan Carlsson. "CNN Features Off-the-Shelf: An Astounding Baseline for Recognition." In: *IEEE Conference on Computer Vision and Pattern Recognition (CVPR)* (Columbus, OH, US, 23.06.–28.06.2014). IEEE Conference on Computer Vision and Pattern Recognition. Piscataway, NJ, US: IEEE, 2014, pp. 512–519. ISBN: 9781479951192. https://doi.org/10.1109/CVPRW.2014.131.

283. Waseem Rawat and Zenghui Wang. "Deep Convolutional Neural Networks for Image Classification: A Comprehensive Review." In: *Neural Computation* 29 (9 2017), pp. 2352–2449. ISSN: 0899-7667. https://doi.org/10.1162/neco_a_00990.
284. Xuning Feng, Caihao Weng, Xiangming He, Xuebing Han, Languang Lu, Dongsheng Ren, and Minggao Ouyang. "Online State-of-Health Estimation for Li-Ion Battery Using Partial Charging Segment Based on Support Vector Machine." In: *IEEE Transactions on Vehicular Technology* 68 (9 2019), pp. 8583–8592. ISSN: 0018-9545. https://doi.org/10.1109/TVT.2019.2927120.
285. Mattin Lucu, Egoitz Martinez-Laserna, Iñigo Gandiaga, and Haritza Camblong. "A critical review on self-adaptive Li-ion battery ageing models." In: *Journal of Power Sources* 401 (2018), pp. 85–101. ISSN: 03787753. https://doi.org/10.1016/j.jpowsour.2018.08.064.
286. Sudipta Bijoy Sarmah, Pankaj Kalita, Akhil Garg, Xiao-dong Niu, Xing-Wei Zhang, Xiongbin Peng, and Dipanwita Bhattacharjee. "A Review of State of Health Estimation of Energy Storage Systems. Challenges and Possible Solutions for Futuristic Applications of Li-Ion Battery Packs in Electric Vehicles." In: *Journal of Electrochemical Energy Conversion and Storage* 16 (4 2019), p. 525. ISSN: 2381-6872. https://doi.org/10.1115/1.4042987.
287. Maitane Berecibar, Iñigo Gandiaga, Igor Villarreal Sarria, Noshin Omar, Joeri van Mierlo, and Peter van den Bossche. "Critical review of state of health estimation methods of Li-ion batteries for real applications." In: *Renewable and Sustainable Energy Reviews* 56 (2016), pp. 572–587. ISSN: 13640321. https://doi.org/10.1016/j.rser.2015.11.042.
288. Muratahan Aykol, Chirranjeevi Balaji Gopal, Abraham Anapolsky, Patrick K. Herring, Bruis van Vlijmen, Marc D. Berliner, Martin Z. Bazant, Richard D. Braatz,William C. Chueh, and Brian D. Storey. "Perspective-Combining Physics and Machine Learning to Predict Battery Lifetime." In: *Journal of The Electrochemical Society* 168 (3 2021), p. 030525. ISSN: 0013-4651. https://doi.org/10.1149/1945-7111/abec55.
289. Huixing Meng and Yan-Fu Li. "A review on prognostics and health management (PHM) methods of lithium-ion batteries." In: *Renewable and Sustainable Energy Reviews* 116 (2019). ISSN: 13640321. https://doi.org/10.1016/j.rser.2019.109405.
290. Kailong Liu, Zhongbao Wei, Chenghui Zhang, Yunlong Shang, Remus Teodorescu, and Qing-Long Han. "Towards Long Lifetime Battery. AI-Based Manufacturing and Management." In: *IEEE/CAA Journal of Automatica Sinica* 9 (7 2022), pp. 1139–1165. ISSN: 2329-9266. https://doi.org/10.1109/JAS.2022.105599.
291. Cambridge University Press. *Cambridge Advanced Learner's Dictionary*. 3rd ed. Cambridge, UK: Cambridge University Press, 2008. 100 pp. ISBN: 978-0521674683.
292. Cambridge University Press. *Cambridge Academic Content Dictionary*. 1st ed. Cambridge, UK: Cambridge University Press, 2009. 1137 pp. ISBN: 9780521691963.
293. Albert Sydney Hornby, Diana Lea, and Jennifer Bradbery. *Oxford advanced learner's dictionary of current English*. 10th ed. 2020. xii, 1820,WT30, ST10, V56, R31 Seiten. ISBN: 978-0194798594.
294. Diana Lea. *Oxford learner's dictionary of academic English*. 1st ed. Oxford, UK: Oxford Univ. Press, 2014. 47 pp. ISBN: 978-0194333504.
295. Angus Stevenson. *The Oxford dictionary of English*. 3rd ed. Oxford, UK: Oxford Univ. Press, 2010. ISBN: 978-0199571123. url: http://www.oxfordreference.com/view/10.1093/acref/9780199571123.001.0001/acref-9780199571123.

296. Stella O'Shea. *Cambridge business english dictionary*. Stuttgart and Cambrigde: Ernst Klett Sprachen GmbH and Cambridge University Press, 2011. ISBN: 978-0521122504.
297. Fotios Petropoulos, Daniele Apiletti, Vassilios Assimakopoulos, Mohamed Zied Babai, Devon K. Barrow, Souhaib Ben Taieb, Christoph Bergmeir, Ricardo J. Bessa, Jakub Bijak, John E. Boylan, Jethro Browell, Claudio Carnevale, Jennifer L. Castle, Pasquale Cirillo, Michael P. Clements, Clara Cordeiro, Fernando Luiz Cyrino Oliveira, Shari de Baets, Alexander Dokumentov, Joanne Ellison, Piotr Fiszeder, Philip Hans Franses, David T. Frazier, Michael Gilliland, M. Sinan Gönül, Paul Goodwin, Luigi Grossi, Yael Grushka-Cockayne, Mariangela Guidolin, Massimo Guidolin, Ulrich Gunter, Xiaojia Guo, Renato Guseo, Nigel Harvey, David F. Hendry, Ross Hollyman, Tim Januschowski, Jooyoung Jeon, Victor Richmond R. Jose, Yanfei Kang, Anne B. Koehler, Stephan Kolassa, Nikolaos Kourentzes, Sonia Leva, Feng Li, Konstantia Litsiou, Spyros Makridakis, Gael M. Martin, Andrew B. Martinez, Sheik Meeran, Theodore Modis, Konstantinos Nikolopoulos, Dilek Önkal, Alessia Paccagnini, Anastasios Panagiotelis, Ioannis Panapakidis, Jose M. Pavía, Manuela Pedio, Diego J. Pedregal, Pierre Pinson, Patrícia Ramos, David E. Rapach, J. James Reade, Bahman Rostami-Tabar, Michał Rubaszek, Georgios Sermpinis, Han Lin Shang, Evangelos Spiliotis, Aris A. Syntetos, Priyanga Dilini Talagala, Thiyanga S. Talagala, Len Tashman, Dimitrios Thomakos, Thordis Thorarinsdottir, Ezio Todini, Juan Ramón Trapero Arenas, Xiaoqian Wang, Robert L. Winkler, Alisa Yusupova, and Florian Ziel. "Forecasting. Theory and practice." In: *International Journal of Forecasting* 38 (3 2022), pp. 705–871. ISSN: 0169-2070. https://doi.org/10.1016/j.ijforecast.2021.11.001.
298. Vidhi Chugh. *Estimation, Prediction and Forecasting. Estimation vs prediction. Prediction vs Forecast. Read along for a quick distinction*. 2020. url: https://towardsdatascience.com/estimation-prediction-and-forecasting-40c56a5be0c9 (visited on 01.03.2023).
299. Matthias Döring. *Prediction vs Forecasting. Predictions do not always concern the future ...* 2018. url: https://www.datascienceblog.net/post/machine-learning/forecasting_vs_prediction (visited on 01.03.2023).
300. Zhen Guo, Xinping Qiu, Guangdong Hou, Boryann Liaw, and Changshui Zhang. "State of health estimation for lithium ion batteries based on charging curves." In: *Journal of Power Sources* 249 (2014), pp. 457–462. ISSN: 03787753. https://doi.org/10.1016/j.jpowsour.2013.10.114.
301. Lingjun Song, Keyao Zhang, Tongyi Liang, Xuebing Han, and Yingjie Zhang. "Intelligent state of health estimation for lithium-ion battery pack based on big data analysis." In: *Journal of Energy Storage* 32 (2020), p. 101836. ISSN: 2352152X. https://doi.org/10.1016/j.est.2020.101836.
302. Duo Yang, Xu Zhang, Rui Pan, Yujie Wang, and Zonghai Chen. "A novel Gaussian process regression model for state-of-health estimation of lithium-ion battery using charging curve." In: *Journal of Power Sources* 384 (1 2018), pp. 387–395. ISSN: 03787753. https://doi.org/10.1016/j.jpowsour.2018.03.015.
303. Huixin Tian, Pengliang Qin, Kun Li, and Zhen Zhao. "Review and recent advances in battery health monitoring and prognostics technologies for electric vehicle (EV) safety and mobility." In: *Journal of Cleaner Production* 261 (2020). ISSN: 0959-6526. https://doi.org/10.1016/j.jclepro.2020.120813.

304. Man-Fai Ng, Jin Zhao, Qingyu Yan, Gareth J. Conduit, and Zhi Wei Seh. "Predicting the state of charge and health of batteries using data-driven machine learning." In: *Nature Machine Intelligence* 2 (3 2020), pp. 161–170. ISSN: 2522-5839. https://doi.org/10.1038/s42256-020-0156-7.
305. Xiaosong Hu, Le Xu, Xianke Lin, and Michael G. Pecht. "Battery Lifetime Prognostics." In: *Joule* 4 (2 2020), pp. 310–346. ISSN: 2542-4351. https://doi.org/10.1016/j.joule.2019.11.018.
306. Yu-Hui Lui, Meng Li, Austin Downey, Sheng Shen, Venkat Pavan Nemani, Hui Ye, Collette van Elzen, Gaurav Jain, Shan Hu, Simon Laflamme, and Chao Hu. "Physics-based prognostics of implantable-grade lithium-ion battery for remaining useful life prediction." In: *Journal of Power Sources* 485 (2021). ISSN: 03787753. https://doi.org/10.1016/j.jpowsour.2020.229327.
307. Liming Deng, Wenjing Shen, Hongfei Wang, and Shuqiang Wang. "A rest-time-based prognostic model for remaining useful life prediction of lithium-ion battery." In: *Neural Computing & Applications* 33 (6 2021), pp. 2035–2046. ISSN: 1433-3058. https://doi.org/10.1007/s00521-020-05105-0.
308. Autovista Group Limited, TWAICE Technologies GmbH, and TÜV Rheinland Mobility. *The Power of Signalling. How Battery Health Reports will support remarketing of used Battery Electric Vehicles (BEVs)*. 2020. url: https://autovistagroup.com/batteryhealth-reports-whitepaper (visited on 01.03.2023).
309. Samuel Greenbank. "Data-driven battery state of health diagnostics and prognostics." Department of Engineering, St. Cross College. PhD Thesis. Oxford, UK, 2022. url: https://ora.ox.ac.uk/objects/uuid:da0cf799-1489-4cb8-9214-4ed5be742275 (visited on 01.03.2023).
310. Thorsten Baumhöfer, Manuel Brühl, Susanne Rothgang, and Dirk Uwe Sauer. "Production caused variation in capacity aging trend and correlation to initial cell performance." In: *Journal of Power Sources* 247 (10 2014), pp. 332–338. ISSN: 03787753. https://doi.org/10.1016/j.jpowsour.2013.08.108.
311. Stephen J. Harris, David J. Harris, and Chen Li. "Failure statistics for commercial lithium ion batteries. A study of 24 pouch cells." In: *Journal of Power Sources* 342 (2017), pp. 589–597. ISSN: 03787753. https://doi.org/10.1016/j.jpowsour.2016.12.083.
312. Hermann G. Matthies. "Quantifying uncertainty: Modern computational representation of probability and applications." In: *Extreme man-made and natural hazards in dynamics of structures*. Ed. by Adnan Ibrahimbegović and Ivica Kozar. NATO security through science series. Dordrecht, The Netherlands: Springer, 2007, pp. 105–136. ISBN: 1402056567. https://doi.org/10.1007/978-1-4020-5656-7_4.
313. Armen Der Kiureghian and Ove Ditlevsen. "Aleatory or epistemic? Does it matter?" In: *Structural Safety* 31 (2 2009), pp. 105–112. ISSN: 01674730. https://doi.org/10.1016/j.strusafe.2008.06.020.
314. Eyke Hüllermeier and Willem Waegeman. "Aleatoric and epistemic uncertainty in machine learning. An introduction to concepts and methods." In: *Machine Learning* 110 (3 2021), pp. 457–506. ISSN: 0885-6125. https://doi.org/10.1007/s10994-021-05946-3.
315. Felix Heinrich. "Battery State Estimation of Electric Vehicles using Neural Networks." Dept. of Computer Science. PhD Thesis. Erlangen, Germany, 2022. 182 pp. url: https://nbn-resolving.org/urn:nbn:de:bvb:29-opus4-191741 (visited on 01.03.2023).

316. Wei He, Nicholas Williard, Michael Osterman, and Michael G. Pecht. "Prognostics of lithium-ion batteries based on Dempster-Shafer theory and the Bayesian Monte Carlo method." In: *Journal of Power Sources* 196 (23 2011), pp. 10314–10321. ISSN: 03787753. https://doi.org/10.1016/j.jpowsour.2011.08.040.
317. Michael E. Porter. *Competitive advantage. Creating and sustaining superior performance* 1st ed. New York, NY, US: Free Press, 1998. 557 pp. ISBN: 0-684-84146-0.
318. Muhammad Amer, Tugrul U. Daim, and Antonie Jetter. "A review of scenario planning." In: *Futures* 46 (February 2013), pp. 23–40. ISSN: 00163287. https://doi.org/10.1016/j.futures.2012.10.003.
319. Ulf Pillkahn. *Using Trends and Scenarios as Tools for Strategy Development*. 1st ed. Hoboken, NJ, US: Wiley-VCH, 2008. 453 pp. ISBN: 978-3-89578-304-3.
320. John S. Pruitt and Tamara Adlin. *The persona lifecycle. Keeping people in mind throughout product design*. The Morgan Kaufmann series in interactive technologies. Amsterdam and Boston: Elsevier, 2006. 724 pp. ISBN: 9780125662512. https://doi.org/10.1016/B978-0-12-566251-2.X5000-X.
321. Alan Cooper. *The inmates are running the asylum. Why high-tech products drive us crazy and how to restore the sanity*. 6th ed. Indianapolis, IN, US: Sams, 2006. 255 pp. ISBN: 978-0672326141.
322. Jan G. de Gooijer and Rob J. Hyndman. "25 years of time series forecasting." In: *International Journal of Forecasting* 22 (3 2006), pp. 443–473. ISSN: 0169-2070. https://doi.org/10.1016/j.ijforecast.2006.01.001.
323. Ling Shao, Fan Zhu, and Xuelong Li. "Transfer learning for visual categorization: a survey." In: IEEE Transactions on Neural Networks and Learning Systems 26 (5 2015), pp. 1019–1034. ISSN: 2162-2388. https://doi.org/10.1109/TNNLS.2014.2330900.
324. Sebastian Ruder, Matthew E. Peters, Swabha Swayamdipta, and Thomas Wolf. "Transfer Learning in Natural Language Processing." In: *Proceedings of the 2019 Conference of the North American Chapter of the Association for Computational Linguistics: Tutorials* (Minneapolis, MN, US, 02.06.2019). Ed. by Anoop Sarkar and Michael Strube. 2019, pp. 15–18. https://doi.org/10.18653/v1/N19-5004.
325. Weihan Li, Neil Sengupta, Philipp André Dechent, David A. Howey, Anuradha Annaswamy, and Dirk Uwe Sauer. "One-shot battery degradation trajectory prediction with deep learning." In: *Journal of Power Sources*8 (1 2021), p. 230024. ISSN: 03787753. https://doi.org/10.1016/j.jpowsour.2021.230024.
326. Sheng Shen, Mohammadkazem Sadoughi, Meng Li, Zhengdao Wang, and Chao Hu. "Deep convolutional neural networks with ensemble learning and transfer learning for capacity estimation of lithium-ion batteries." In: *Applied Energy* 260 (9 2020), p. 114296. ISSN: 03062619; 1872-9118. https://doi.org/10.1016/j.apenergy.2019.114296.
327. Minho Kim and Soohee Han. "Novel Data-Efficient Mechanism-Agnostic Capacity Fade Model for Li-Ion Batteries." In: *IEEE Transactions on Industrial Electronics* 68 (7 2021), pp. 6267–6275. ISSN: 0278-0046. https://doi.org/10.1109/TIE.2020.2996156.
328. Elixabet Sarasketa-Zabala, Egoitz Martinez-Laserna, Maitane Berecibar, Iñigo Gandiaga, L. M. Rodriguez-Martinez, and Igor Villarreal Sarria. "Realistic lifetime prediction approach for Li-ion batteries." In: *Applied Energy* 162 (2016), pp. 839–852. ISSN: 03062619; 1872-9118. https://doi.org/10.1016/j.apenergy.2015.10.115.

329. Robert R. Richardson, Michael A. Osborne, and David A. Howey. "Gaussian process regression for forecasting battery state of health." In: *Journal of Power Sources* 357 (2017), pp. 209–219. ISSN: 03787753. https://doi.org/10.1016/j.jpowsour.2017.05.004.
330. Jing Lin, Yu Zhang, and Edwin Khoo. "Hybrid physics-based and data-driven modeling with calibrated uncertainty for lithium-ion battery degradation diagnosis and prognosis." In: *Tackling Climate Change with Machine Learning: workshop at NeurIPS 2021* (2021). https://doi.org/10.48550/arXiv.2110.13661.
331. Jose Genario Jr de Oliveira, Vipul Dhingra, and Christoph Hametner. "Feature Extraction, Ageing Modelling and Information Analysis of a Large-Scale Battery Ageing Experiment." In: *Energies* 14 (17 2021). ISSN: 1996-1073. https://doi.org/10.3390/en14175295.
332. Adnan Nuhic, Jonas Bergdolt, Bernd Spier, Michael Buchholz, and Klaus Dietmayer. "Battery Health Monitoring and Degradation Prognosis in Fleet Management Systems." In: *World Electric Vehicle Journal* 9 (3 2018), p. 39. ISSN: 2032-6653. https://doi.org/10.3390/wevj9030039.
333. Viraj Phatarphod. "Adapting a data-driven battery ageing model to make remaininguseful-life estimations using dynamic vehicle data." School of Engineering Sciences in Chemistry, Biotechnology and Health. Degree Project. Stockholm, Sweden, 2021. url: http://urn.kb.se/resolve?urn=urn:nbn:se:kth:diva-305828.
334. Andreas Udo Sass, Enes Esatbeyoğlu, and Till Iwwerks. "Data-Driven Powertrain Component Aging Prediction Using In-Vehicle Signals." In: *SOFSEM (Doctoral Student Research Forum)* (2020). url: http://ceur-ws.org/Vol-2568/paper10.pdf.
335. Caifeng Zheng, Weirong Liu, Bin Chen, Dianzhu Gao, Yijun Cheng, Yingze Yang, Xiaoyong Zhang, Shuo Li, Zhiwu Huang, and Jun Peng. "A Data-driven Approach for Remaining Useful Life Prediction of Aircraft Engines." In: *21st International Conference on Intelligent Transportation Systems (ITSC)*. 2018 21st International Conference on Intelligent Transportation Systems (ITSC) (Maui, HI, US, 04.11.–07.11.2018). IEEE, 2018, pp. 184–189. ISBN: 978-1-7281-0321-1. https://doi.org/10.1109/ITSC.2018.8569915.
336. Kar Hoou Hui, Ching Sheng Ooi, Meng Hee Lim, Mohd Salman Leong, and Salah Mahdi Al-Obaidi. "An improved wrapper-based feature selection method for machinery fault diagnosis." *In: PloS one* 12 (12 2017), e0189143. https://doi.org/10.1371/journal.pone.0189143.
337. Felix Richter, Oliver Hartkopp, and Dirk C. Mattfeld. "Automatic Defect Detection by Classifying Aggregated Vehicular Behavior." In: *Foundations of intelligent systems. 23rd International Symposium, ISMIS 2017* (Warsaw, Poland, 26.06.–29.06.2017). Ed. by Marzena Kryszkiewicz, Annalisa Appice, Dominik Slezak, Henryk Rybiński, Andrzej Skowroński, and Zbigniew W. Raś. Lecture Notes in Computer Science 10352. International Symposium on Methodologies for Intelligent Systems. Cham, Switzerland: Springer, 2017, pp. 205–214. ISBN: 978-3-319-60438-1. https://doi.org/10.1007/978-3-319-60438-1_21.
338. Hongxun Liu, Yangyang Wang, Wenhua Li, Fangxu Shao, and Mingze He. "Decay mechanism and capacity prediction of lithium-ion batteries under low-temperature near-adiabatic condition." In: *Inorganic chemistry communications* 137 (2022). ISSN: 1879-0259. https://doi.org/10.1016/j.inoche.2021.109151.

339. Kailong Liu, T. R. Ashwin, Xiaosong Hu, Mattin Lucu, and Widanalage Dhammika Widanage. "An evaluation study of different modelling techniques for calendar ageing prediction of lithium-ion batteries." In: *Renewable and Sustainable Energy Reviews* 131 (6273 2020), p. 110017. ISSN: 13640321. https://doi.org/10.1016/j.rser.2020.110017.
340. Jin-zhen Kong, Dong Wang, Tongtong Yan, Jingzhe Zhu, and Xi Zhang. "Accelerated Stress Factors Based Nonlinear Wiener Process Model for Lithium-Ion Battery Prognostics." In: *IEEE Transactions on Industrial Electronics* 69 (11 2022), pp. 11665–11674. ISSN: 0278-0046. https://doi.org/10.1109/TIE.2021.3127035.
341. Muhammad Umair Ali, Karam Dad Kallu, Haris Masood, Kamran Ali Khan Niazi, Muhammad Junaid Alvi, Usman Ghafoor, and Amad Zafar. "Kernel recursive least square tracker and long-short term memory ensemble based battery health prognostic model." In: *iScience* 24 (11 2021). https://doi.org/10.1016/j.isci.2021.103286.
342. Lisen Yan, Jun Peng, Dianzhu Gao, Yue Wu, Yongjie Liu, Heng Li, Weirong Liu, and Zhiwu Huang. "A hybrid method with cascaded structure for early-stage remaining useful life prediction of lithium-ion battery." In: *Energy* 243 (5 2022), p. 123038. ISSN: 1873-6785. https://doi.org/10.1016/j.energy.2021.123038.
343. Logan Ward, Susan Babinec, Eric J. Dufek, David A. Howey, Venkatasubramanian Viswanathan, Muratahan Aykol, David A.C. Beck, Benjamin Blaiszik, Bor-Rong Chen, George Crabtree, Simon Clark, Valerio de Angelis, Philipp André Dechent, Matthieu Dubarry, Erica E. Eggleton, Donal P. Finegan, Ian Foster, Chirranjeevi Balaji Gopal, Patrick K. Herring, Victor W. Hu, Noah H. Paulson, Yuliya Preger, Dirk Uwe Sauer, Kandler Smith, Seth W. Snyder, Shashank Sripad, Tanvir R. Tanim, and Linnette Teo. "Principles of the Battery Data Genome." In: *Joule* 6 (10 2022), pp. 2253–2271. ISSN: 2542-4351. https://doi.org/10.1016/j.joule.2022.08.008.
344. Dirk Uwe Sauer and Heinz Wenzl. "BATTERIES | Lifetime Prediction." In: *Encyclopedia of electrochemical power sources*. Ed. by Jürgen Garche. 1st ed. Amsterdam, The Netherlands: Elsevier, 2009, pp. 522–538. ISBN: 978-0-444-52745-5. https://doi.org/10.1016/B978-044452745-5.00852-2.
345. Shuo Sun, Junzhong Sun, Zongliang Wang, Zhiyong Zhou, and Wei Cai. "Prediction of Battery SOH by CNN-BiLSTM Network Fused with Attention Mechanism." In: *Energies* 15 (12 2022), p. 4428. ISSN: 1996-1073. https://doi.org/10.3390/en15124428.
346. Kailong Liu, Xiaosong Hu, Zhongbao B. Wei, Yi Li, and Yan Jiang. "Modified Gaussian Process Regression Models for Cyclic Capacity Prediction of Lithium-Ion Batteries." In: *IEEE Transactions on Transportation Electrification* 5 (4 2019), pp. 1225–1236. ISSN: 2332-7782. https://doi.org/10.1109/TTE.2019.2944802.
347. Pasquale de Falco, Luigi Pio Di Noia, and Renato Rizzo. "State of Health Prediction of Lithium-Ion Batteries Using Accelerated Degradation Test Data." In: *IEEE Transactions on Industry Applications* 57 (6 2021), pp. 6483–6493. ISSN: 1939-9367. https://doi.org/10.1109/TIA.2021.3112392.
348. Karl-Heinz Pettinger, Achim Kampker, Claus-Rupert Hohenthanner, Christoph Deutskens, Heiner Hans Heimes, and Ansgar vom Hemdt. "Lithium-ion cell and battery production processes." In: *Lithium-Ion Batteries: Basics and Applications*. Ed. by Reiner Korthauer. Berlin, Heidelberg, Germany: Springer Berlin Heidelberg, 2018, pp. 211–226. ISBN: 978-3-662-53069-6. https://doi.org/10.1007/978-3-662-53071-9_17.
349. Lei Yao, Shiming Xu, Aihua Tang, Fang Zhou, Junjian Hou, Yanqiu Xiao, and Zhijun Fu. "A Review of Lithium-Ion Battery State of Health Estimation and Prediction Methods."

In: *World Electric Vehicle Journal* 12 (3 2021), p. 113. ISSN: 2032-6653. https://doi.org/10.3390/wevj12030113.
350. Shunli Wang, Siyu Jin, Dekui Bai, Yongcun Fan, Haotian Shi, and Carlos Fernandez. "A critical review of improved deep learning methods for the remaining useful life prediction of lithium-ion batteries." In: *Energy Reports* 7 (2021), pp. 5562–5574. ISSN: 2352-4847. https://doi.org/10.1016/j.egyr.2021.08.182.
351. Lifeng Wu, Xiaohui Fu, and Yong Guan. "Review of the Remaining Useful Life Prognostics of Vehicle Lithium-Ion Batteries Using Data-Driven Methodologies." In: *Applied Sciences* 6 (6 2016), p. 166. ISSN: 2076-3417. https://doi.org/10.3390/app6060166.
352. Nils Collath, Benedikt Tepe, Stefan Englberger, Andreas Jossen, and Holger Hesse. "Aging aware operation of lithium-ion battery energy storage systems. A review." In: *Journal of Energy Storage* 55 (11 2022), p. 105634. ISSN: 2352152X. https://doi.org/10.1016/j.est.2022.105634.
353. Jane Webster and Richard T. Watson. "Analyzing the Past to Prepare for the Future: Writing a literature Review." In: *MIS Quarterly* 26 (2 2002), pp. xiii–xxiii. url: https://www.jstor.org/stable/4132319
354. Clarivate. Web of Science. 2021. url: https://www.webofscience.com (visited on 01.03.2023).
355. Google. Google Scholar. 2021. url: https://scholar.google.com (visited on 01.03.2023).
356. Martin Boeker, Werner Vach, and Edith Motschall. "Google Scholar as replacement for systematic literature searches. Good relative recall and precision are not enough." In: *BMC medical research methodology* 13 (2013), p. 131. https://doi.org/10.1186/1471-2288-13-131.
357. The Allen Institute for Artificial Intelligence. Semantic Scholar. 2021. url: https://www.semanticscholar.org (visited on 01.03.2023).
358. Universitätsbibliothek Bielefeld. *Bielefeld Academic Search Engine* (BASE). 2021. url: https://www.base-search.net (visited on 01.03.2023).
359. Jan vom Brocke, Alexander Simons, Björn Niehaves, Kai Riemer, Ralf Plattfaut, and Anne Cleven. "Reconstructing the Giant: On the Importance of Rigour in Documenting the Literature Search Process." In: *European Conference on Information Systems (ECIS) 2009 Proceedings* (2009). url: http://aisel.aisnet.org/ecis2009/161.
360. Bolun Xu, Alexandre Oudalov, Andreas Ulbig, Goran Andersson, and Daniel S. Kirschen. "Modeling of Lithium-Ion Battery Degradation for Cell Life Assessment." In: *IEEE Transactions on Smart Grid* 9 (2 2018), pp. 1131–1140. ISSN: 1949-3061. https://doi.org/10.1109/TSG.2016.2578950.
361. Dong Wang, Jin-zhen Kong, Fangfang Yang, Yang Zhao, and Kwok-Leung Tsui. "Battery prognostics at different operating conditions." In: *Meausrement* 151 (15 2020), p. 107182. ISSN: 02632241. https://doi.org/10.1016/j.measurement.2019.107182.
362. Eduardo Redondo-Iglesias, Pascal Venet, and Serge Pelissier. "Modelling Lithium- Ion Battery Ageing in Electric Vehicle Applications-Calendar and Cycling Ageing Combination Effects." In: *Batteries* 6 (1 2020). ISSN: 2313-0105. https://doi.org/10.3390/batteries6010014.
363. Andrea Carolina Cordoba-Arenas, Simona Onori, Yann Guezennec, and Giorgio Rizzoni. "Capacity and power fade cycle-life model for plug-in hybrid electric vehicle lithium-ion battery cells containing blended spinel and layered-oxide positive electro-

des." In: *Journal of Power Sources* 278 (2015), pp. 473–483. ISSN: 03787753. https://doi.org/10.1016/j.jpowsour.2014.12.047.

364. Maik Naumann, Franz B. Spingler, and Andreas Jossen. "Analysis and modeling of cycle aging of a commercial LiFePO4/graphite cell." In: *Journal of Power Sources* 451 (2020). ISSN: 03787753. https://doi.org/10.1016/j.jpowsour.2019.227666.
365. Weijun Gu, Zechang Sun, Xuezhe Wei, and Haifeng Dai. "A Capacity Fading Model of Lithium-Ion Battery Cycle Life Based on the Kinetics of Side Reactions for Electric Vehicle Applications." In: *Electrochimica ACTA* 133 (2014), pp. 107–116. ISSN: 1873-3859. https://doi.org/10.1016/j.electacta.2014.03.186.
366. Josu Olmos, Iñigo Gandiaga, Andoni Saez-de-Ibarra, Xabier Larrea, Txomin Nieva, and Iosu Aizpuru. "Modelling the cycling degradation of Li-ion batteries: Chemistry influenced stress factors." In: *Journal of Energy Storage* 40 (2021). ISSN: 2352152X. https://doi.org/10.1016/j.est.2021.102765.
367. Li Chen, Yuqi Q. Tong, and Zuomin M. Dong. "Li-Ion Battery Performance Degradation Modeling for the Optimal Design and Energy Management of Electrified Propulsion Systems." In: *Energies* 13 (7 2020). ISSN: 1996-1073. https://doi.org/10.3390/en13071629.
368. Andrew Chu, Anirudh Allam, Andrea Carolina Cordoba-Arenas, Giorgio Rizzoni, and Simona Onori. "Stochastic capacity loss and remaining useful life models for lithium-ion batteries in plug-in hybrid electric vehicles." In: *Journal of Power Sources* 478 (2020). ISSN: 03787753. https://doi.org/10.1016/j.jpowsour.2020.228991.
369. Joris de Hoog, Jean-Marc Timmermans, Daniel Ioan-Stroe, Maciej Jozef Swierczynski, Joris Jaguemont, Shovon Goutam, Noshin Omar, Joeri van Mierlo, and Peter van den Bossche. "Combined cycling and calendar capacity fade modeling of a Nickel-Manganese-Cobalt Oxide Cell with real-life profile validation." In: *Applied Energy* 200(2017), pp. 47–61. ISSN: 03062619; 1872-9118. https://doi.org/10.1016/j.apenergy.2017.05.018.
370. Paul Gasper, Kevin L. Gering, Eric J. Dufek, and Kandler Smith. "Challenging Practices of Algebraic Battery Life Models through Statistical Validation and Model Identification via Machine-Learning." In: *Journal of The Electrochemical Society* 168 (2 2021). ISSN: 0013-4651. https://doi.org/10.1149/1945-7111/abdde1.
371. Mattin Lucu, Markel Azkue, Haritza Camblong, and Egoitz Martinez-Laserna. "Data-Driven Nonparametric Li-Ion Battery Ageing Model Aiming At Learning From Real Operation Data: Holistic Validation With Ev Driving Profiles." In: *ECCE 2020. IEEE Energy Conversion Congress & Expo* (Detroit, MI, US, 11.10.–15.10.2020). Piscataway, NJ, US: IEEE, 2020, pp. 5600-5607. ISBN: 9781728158273. https://doi.org/10.1109/ECCE44975.2020.9235814.
372. Jaouher Ben Ali, Chaima Azizi, Lotfi Saidi, Eric Bechhoefer, and Mohamed Benbouzid. "Reliable state of health condition monitoring of Li-ion batteries based on incremental support vector regression with parameters optimization." In: *Proceedings of the Institution of Mechanical Engineers, Part I: Journal of Systems and Control Engineering* (2020), p. 095965182095084. ISSN: 2041-3041. https://doi.org/10.1177/0959651820950849.
373. Jorn M. Reniers, Grietus Mulder, and David A. Howey. "Review and Performance Comparison of Mechanical-Chemical Degradation Models for Lithium-Ion Batteries." In: *Journal of The Electrochemical Society* 166 (14 2019), A3189-A3200. ISSN: 0013-4651. https://doi.org/10.1149/2.0281914jes.

374. M. Sanjeev Arulampalam, Simon Maskell, Neil Gordon, and Tim C. Clapp. "A tutorial on particle filters for online nonlinear/non-Gaussian Bayesian tracking." In: *IEEE Transactions on Signal Processing* 50 (2 2002), pp. 174–188. ISSN: 1053587X. https://doi.org/10.1109/78.978374.
375. Datong Liu, Jingyue Pang, Jianbao Zhou, Yu Peng, and Michael G. Pecht. "Prognostics for state of health estimation of lithium-ion batteries based on combination Gaussian process functional regression." In: *Microelectronics Reliability* 53 (6 2013), pp. 832–839. ISSN: 00262714. https://doi.org/10.1016/j.microrel.2013.03.010.
376. Francesco Cadini, Claudio Sbarufatti, Francesco Cancelliere, and Marco Giglio. "State-of-life prognosis and diagnosis of lithium-ion batteries by data-driven particle filters." In: *Applied Energy* 235 (2019), pp. 661–672. ISSN: 03062619; 1872-9118. https://doi.org/10.1016/j.apenergy.2018.10.095.
377. Yang Chang and Huajing Fang. "A hybrid prognostic method for system degradation based on particle filter and relevance vector machine." In: *Reliability Engineering and System Safety* 186 (2019), pp. 51–63. ISSN: 1879-0836. https://doi.org/10.1016/j.ress.2019.02.011.
378. Dong Wang, Jin-zhen Kong, Yang Zhao, and Kwok-Leung Tsui. "Piecewise model based intelligent prognostics for state of health prediction of rechargeable batteries with capacity regeneration phenomena." In: *Meausrement* 147 (2019). ISSN: 02632241. https://doi.org/10.1016/j.measurement.2019.07.064.
379. Seongyoon Kim, Yun Young Choi, Ki Jae Kim, and Juni-Il Choi. "Forecasting state-of-health of lithium-ion batteries using variational long short-term memory with transfer learning." In: *Journal of Energy Storage* 41 (2021). ISSN: 2352152X. https://doi.org/10.1016/j.est.2021.102893.
380. Kailong Liu, Yi Li, Xiaosong Hu, Mattin Lucu, and Widanalage Dhammika Widanage. "Gaussian Process Regression With Automatic Relevance Determination Kernel for Calendar Aging Prediction of Lithium-Ion Batteries." In: *IEEE Transactions on Industrial Informatics* 16 (6 2020), pp. 3767–3777. ISSN: 1551-3203. https://doi.org/10.1109/TII.2019.2941747.
381. Weihan Li, Haotian Zhang, Bruis van Vlijmen, Philipp André Dechent, and Dirk Uwe Sauer. "Forecasting battery capacity and power degradation with multi-task learning." In: *Energy Storage Materials* 53 (3 2022), pp. 453–466. ISSN: 24058297. https://doi.org/10.1016/j.ensm.2022.09.013.
382. Seàn McLoone and George Irwin. "Improving neural network training solutions using regularisation." In: *Neurocomputing*37 (1–4 2001), pp. 71–90. ISSN: 09252312. https://doi.org/10.1016/S0925-2312(00)00314-3.
383. Antti Aitio and David A. Howey. "Predicting battery end of life from solar off-grid system field data using machine learning." In: *Joule* 5 (12 2021), pp. 3204–3220. ISSN: 2542-4351. https://doi.org/10.1016/j.joule.2021.11.006.
384. Gonçalo dos Reis, Calum Strange, Mohit Yadav, and Shawn Li. "Lithium-ion battery data and where to find it." In: *Energy and AI* 5 (4 2021), p. 100081. ISSN: 26665468. https://doi.org/10.1016/j.egyai.2021.100081.
385. Falak Naaz, Aniruddh Herle, Janamejaya Channegowda, Aditya Raj, and Meenakshi Lakshminarayanan. "A generative adversarial network-based synthetic data augmentation technique for battery condition evaluation." In: *International Journal of Energy*

Research 45 (13 2021), pp. 19120–19135. ISSN: 1099-114X. https://doi.org/10.1002/er.7013.

386. Jingyuan Zhao and Andrew F. Burke. "Electric Vehicle Batteries. Status and Perspectives of Data-Driven Diagnosis and Prognosis." In: *Batteries*8 (10 2022), p. 142. ISSN: 2313-0105. https://doi.org/10.3390/batteries8100142.
387. Sunday Ochella, Mahmood Shafiee, and Fateme Dinmohammadi. "Artificial intelligence in prognostics and health management of engineering systems." In: *Engineering Applications of Artificial Intelligence* 108 (2022). ISSN: 09521976. https://doi.org/10.1016/j.engappai.2021.104552.
388. Xiaopeng Tang, Kailong Liu, Xin Wang, Furong Gao, James Macro, and Widanalage Dhammika Widanage. "Model Migration Neural Network for Predicting Battery Aging Trajectories." In: *IEEE Transactions on Transportation Electrification* 6 (2 2020), pp. 363–374. ISSN: 2332-7782. https://doi.org/10.1109/TTE.2020.2979547.
389. Jian Guo, Zhaojun J. Li, and Michael G. Pecht. "A Bayesian approach for Li-Ion battery capacity fade modeling and cycles to failure prognostics." In: *Journal of Power Sources* 281 (2015), pp. 173–184. ISSN: 03787753. https://doi.org/10.1016/j.jpowsour.2015.01.164.
390. Mattin Lucu, Egoitz Martinez-Laserna, Iñigo Gandiaga, Kailong Liu, Haritza Camblong, Widanalage Dhammika Widanage, and James Marco. "Data-driven nonparametric Li-ion battery ageing model aiming at learning from real operation data—Part B. Cycling operation." In: *Journal of Energy Storage* 30 (2020), p. 101410. ISSN: 2352152X. https://doi.org/10.1016/j.est.2020.101410.
391. Dave Andre, Christian Appel, Thomas Soczka-Guth, and Dirk Uwe Sauer. "Advanced mathematical methods of SOC and SOH estimation for lithium-ion batteries." In: *Journal of Power Sources* 224 (2013), pp. 20–27. ISSN: 03787753. https://doi.org/10.1016/j.jpowsour.2012.10.001.
392. Robert R. Richardson, Michael A. Osborne, and David A. Howey. "Battery health prediction under generalized conditions using a Gaussian process transition model." In: *Journal of Energy Storage* 23 (2019), pp. 320–328. ISSN: 2352152X. https://doi.org/10.1016/j.est.2019.03.022.
393. Stephan Rohr, Stefan Müller, Michael Baumann, Matthias Kerler, Fabian Ebert, D. Kaden, and Markus Lienkamp. "Quantifying Uncertainties in Reusing Lithium-Ion Batteries from Electric Vehicles." In: *Procedia Manufacturing* 8 (3 2017), pp. 603–610. ISSN: 23519789. https://doi.org/10.1016/j.promfg.2017.02.077.
394. Mohammadhosein Safari, Mathieu Morcrette, Anna Teyssot, and Charles Delacourt. "Life Prediction Methods for Lithium-Ion Batteries Derived from a Fatigue Approach. Approach: II. Capacity-Loss Prediction of Batteries Subjected to Complex Current Profiles." In: *Journal of The Electrochemical Society* 157 (7 2010), A892. ISSN: 0013-4651. https://doi.org/10.1149/1.3432560.
395. Tatsuo Endo, Koichi Mitsunaga, Kiyohum Takahashi, Kakuichi Kobayashi, and Masanori Matsuishi. "Damage evaluation of metals for random or varying loading-three aspects of rain flow method." In: *Proceedings of the 1974 Symposium on Mechanical Behavior of Materials* 1 (1974), pp. 371–380.
396. Jonny Dambrowski, Simon Pichlmaier, and Andreas Jossen. "Mathematical methods for classification of state-of-charge time series for cycle lifetime prediction." In: *Advanced Automotive Batteries Conference Europe (AABC)* (Mainz,

Germany, 18.06.– 22.06.2012). 2012. url: https://www.advancedautobat.com/conferences/previous/AABC-Europe-2012/battery-technology-posters.html.

397. Adnan Nuhic, Ruoyi Hu, Michael Buchholz, and Klaus Dietmayer. "A healthmonitoring and life-prediction approach for lithium-ion batteries based on fatigue analysis." In: *Advanced Automotive Batteries Conference Europe (AABC)* (Strasbourg, France, 24.06.–28.06.2013). 2013. url: https://www.advancedautobat.com/conferences/previous/AABC-Europe-2013/battery-technology-posters.html.
398. Daniel-Ioan Stroe. "Lifetime Models for Lithium-ion Batteries used in Virtual Power Plant Applications." Faculty of Engineering, Science, and Medicine. PhD Thesis. Aalborg, Denmark, 2014. url: https://vbn.aau.dk/da/publications/d56e6622-f1bd-4a88-b82f-f42e6a81e86a.
399. Arvid G. Palmgren. "Die Lebensdauer von Kugellagern." In: *Zeitschrift des Vereines Deutscher Ingenieure* 68 (14 1924), pp. 339–341.
400. Milton A. Miner. "Cumulative Damage in Fatigue." In: *Journal of Applied Mechanics* 12 (3 1945), A159-A164. ISSN: 0021-8936. https://doi.org/10.1115/1.4009458.
401. Erik Frisk, Mattias Krysander, and Emil Larsson. "Data-Driven Lead-Acid Battery Prognostics Using Random Survival Forests." In: *Annual Conference of the Prognostics and Health Management Society* 6 (1 2014). https://doi.org/10.36001/phmconf.2014.v6i1.2370.
402. Yizhou Zhang, Torsten Wik, John Bergström, Michael G. Pecht, and Changfu Zou. "A machine learning-based framework for online prediction of battery ageing trajectory and lifetime using histogram data." In: *Journal of Power Sources* 526 (1 2022), p. 231110. ISSN: 03787753. https://doi.org/10.1016/j.jpowsour.2022.231110.
403. Samuel Greenbank and David A. Howey. "Piecewise-linear modelling with automated feature selection for Li-ion battery end-of-life prognosis." In: *Mechanical Systems and Signal Processing* 184 (2023). ISSN: 0888-3270. https://doi.org/10.1016/j.ymssp.2022.109612.
404. Pongsakorn Chanchaipo and Leelawadee Sirikul. "EV battery degradation forecasting based on high dimensional data." Institutionen för data och informationsteknik. Master Thesis. Göteborg, Sweden, 2020. url: https://hdl.handle.net/20.500.12380/302222.
405. Renato Giorgiani do Nascimento, Matteo Corbetta, Chetan S. Kulkarni, and Felipe A. C. Viana. "Hybrid physics-informed neural networks for lithium-ion battery modeling and prognosis." In: *Journal of Power Sources* 513 (2021). ISSN: 03787753. https://doi.org/10.1016/j.jpowsour.2021.230526.
406. Joonki Hong, Dongheon Lee, Eui-Rim Jeong, and Yung Yi. "Towards the swift prediction of the remaining useful life of lithium-ion batteries with end-to-end deep learning." In: *Applied Energy* 278 (2020). ISSN: 03062619; 1872-9118. https://doi.org/10.1016/j.apenergy.2020.115646.
407. Peter M. Attia, Kristen A. Severson, and Jeremy D. Witmer. "Statistical Learning for Accurate and Interpretable Battery Lifetime Prediction." In: *Journal of The Electrochemical Society* 168 (9 2021), p. 090547. ISSN: 0013-4651. https://doi.org/10.1149/1945-7111/ac2704.
408. Renato Giorgiani do Nascimento. "Lithium-ion Battery Prognosis with Variational Hybrid Physics-informed Neural Networks." Mechanical and Aerospace Engineering. PhD Thesis. Orlando, FL, US, 2022. url: https://stars.library.ucf.edu/etd2020/1011 (visited on 01.03.2023).

409. Fu-Kwun Wang, Chang-Yi Huang, and Tadele Mamo. "Ensemble Model Based on Stacked Long Short-Term Memory Model for Cycle Life Prediction of Lithium- Ion Batteries." In: *Applied Sciences* 10 (10 2020), p. 3549. ISSN: 2076-3417. https://doi.org/10.3390/app10103549.
410. Aramis Perez, Francisco Jaramillo, Vanessa Quintero, and Marcos E. Orchard. "Characterizing the degradation process of Lithium-Ion Batteries using a Similarity- Based-Modeling Approach." In: *Fourth European Conference of the Prognostics and Health Management Society 2018* (2019). https://doi.org/10.36001/phme.2018.v4i1.439.
411. Gang Ning, Bala Haran, and Branko N. Popov. "Capacity fade study of lithium-ion batteries cycled at high discharge rates." In: *Journal of Power Sources* 117 (1-2 2003), pp. 160–169. ISSN: 03787753. https://doi.org/10.1016/S0378-7753(03)00029-6.
412. Kailong Liu, Qiao Peng, Hongbin Sun, Minrui Fei, Huimin Ma, and Tianyu Hu. "A Transferred Recurrent Neural Network for Battery Calendar Health Prognostics of Energy-Transportation Systems." In: *IEEE Transactions on Industrial Informatics* (2022), p. 1. ISSN: 1551-3203. https://doi.org/10.1109/TII.2022.3145573.
413. Clara Bertinelli Salucci, Azzeddine Bakdi, Ingrid Kristine Glad, Erik Vanem, and Riccardo de Bin. "Multivariable Fractional Polynomials for lithium-ion batteries degradation models under dynamic conditions." In: *Journal of Energy Storage* 52 (1 2022), p. 104903. ISSN: 2352152X. https://doi.org/10.1016/j.est.2022.104903.
414. Md Sazzad Hosen, Rekabra Youssef, Theodoros Kalogiannis, Joeri van Mierlo, and Maitane Berecibar. "Battery cycle life study through relaxation and forecasting the lifetime via machine learning." In: *Journal of Energy Storage* 40 (2021), p. 102726. ISSN: 2352152X. https://doi.org/10.1016/j.est.2021.102726.
415. Edward Victor Thomas, Ira Bloom, Jon P. Christophersen, and Vincent S. Battaglia. "Rate-based degradation modeling of lithium-ion cells." In: *Journal of Power Sources* 206 (2012), pp. 378–382. ISSN: 03787753. https://doi.org/10.1016/j.jpowsour.2012.01.106.
416. Edward Victor Thomas, Ira Bloom, Jon P. Christophersen, David C. Robertson, Lee K. Walker, Chinh D. Ho, and Vincent S. Battaglia. "Modeling memoryless degradation under variable stress." In: *Journal of Quality Technology* 51 (3 2019), pp. 284–299. ISSN: 2575-6230. https://doi.org/10.1080/00224065.2019.1569963.
417. Safieh Bamati and Hicham Chaoui. "Lithium-Ion Batteries Long Horizon Health Prognostic Using Machine Learning." In: *IEEE Transactions on Energy Conversion* 37 (2 2022), pp. 1176–1186. ISSN: 1558-0059. https://doi.org/10.1109/TEC.2021.3111525.
418. Christoph Birkl and David A. Howey. *Oxford Battery Degradation Dataset 1*. University of Oxford. 2017. https://doi.org/10.5287/bodleian:KO2kdmYGg. (visited on 01.03.2023).
419. Elio Chiodo, Pasquale de Falco, and Luigi Pio Di Noia. "Probabilistic Modeling of Li-Ion Battery Remaining Useful Life." In: *IEEE Transactions on Industry Applications* 58 (4 2022), pp. 5214–5226. ISSN: 1939-9367. https://doi.org/10.1109/TIA.2022.3170525.
420. Penelope K. Jones, Ulrich Stimming, and Alpha A. Lee. *Impedance-based forecasting of battery performance amid uneven usage*. 2021. https://doi.org/10.5281/zenodo.6645536. (visited on 01.03.2023).
421. Juan Antonio López-Villanueva, Pablo Rodríguez Iturriaga, and Salvador Rodríguez-Bolívar. "A fractional-order model for calendar aging with dynamic storage conditions."

In: *Journal of Energy Storage* 50 (1 2022), p. 104537. ISSN: 2352152X. https://doi.org/10.1016/j.est.2022.104537.

422. Jiahuan Lu, Rui Xiong, Jinpeng Tian, Chenxu Wang, Chia-Wei Hsu, Nien-Ti Tsou, Fengchun Sun, and Ju Li. "Battery degradation prediction against uncertain future conditions with recurrent neural network enabled deep learning." In: *Energy Storage Materials* 50 (2022), pp. 139–151. ISSN: 24058297. https://doi.org/10.1016/j.ensm.2022.05.007.
423. Jiahuan Lu, Rui Xiong, Jinpeng Tian, Chenxu Wang, Chia-Wei Hsu, Nien-Ti Tsou, Fengchun Sun, and Ju Li. *Battery Degradation Dataset (Fixed Current Profiles & Arbitrary Uses Profiles)*. 2022. https://doi.org/10.17632/kw34hhw7xg.3. (visited on 01.03.2023).
424. Junchuan Shi, Alesix. Rivera, and Dazhong Wu. "Battery health management using physics-informed machine learning. Online degradation modeling and remaining useful life prediction." In: *Mechanical Systems and Signal Processing* 179 (2022). ISSN: 0888-3270. https://doi.org/10.1016/j.ymssp.2022.109347.
425. Jianxiang Wang and Yong Xiang. "Fast Modeling of the Capacity Degradation of Lithium-Ion Batteries via a Conditional Temporal Convolutional Encoder-Decoder." In: *IEEE Transactions on Transportation Electrification* 8 (2 2022), pp. 1695–1709. ISSN: 2332-7782. https://doi.org/10.1109/TTE.2021.3128018.
426. Saurabh Saxena, Christopher Hendricks, and Michael G. Pecht. "Cycle life testing and modeling of graphite/LiCoO2 cells under different state of charge ranges." In: *Journal of Power Sources* 327 (5 2016), pp. 394–400. ISSN: 03787753. https://doi.org/10.1016/j.jpowsour.2016.07.057.
427. Heather omdrtz nsöml M. Barkholtz, Armando Fresquez, Babu R. Chalamala, and Summer R. Ferreira. "A Database for Comparative Electrochemical Performance of Commercial 18650-Format Lithium-Ion Cells." In: *Journal of The Electrochemical Society* 164 (12 2017), A2697-A2706. ISSN: 0013-4651. https://doi.org/10.1149/2.1701712jes.
428. Xiaodong Xu, Shengjin Tang, Chuanqiang Yu, Jian Xie, Xuebing Han, and Minggao Ouyang. "Remaining Useful Life Prediction of Lithium-ion Batteries Based onWiener Process Under Time-Varying Temperature Condition." In: *Reliability Engineering and System Safety* 214 (2021). ISSN: 1879-0836. https://doi.org/10.1016/j.ress.2021.107675.
429. Yunwei Zhang, Qiaochu Tang, Yao Zhang, Jiabin Wang, Ulrich Stimming, and Alpha A. Lee. *Identifying degradation patterns of lithium ion batteries from impedance spectroscopy using machine learning*. 2020. https://doi.org/10.5281/zenodo.3633835. (visited on 01.03.2023).
430. Markel Azkue, Mattin Lucu, Egoitz Martinez-Laserna, and Iosu Aizpuru. "Calendar Ageing Model for Li-Ion Batteries Using Transfer Learning Methods." In: *World Electric Vehicle Journal* 12 (3 2021), p. 145. ISSN: 2032-6653. https://doi.org/10.3390/wevj12030145.
431. Yandan Tan and Guangcai Zhao. "Transfer Learning With Long Short-Term Memory Network for State-of-Health Prediction of Lithium-Ion Batteries." In: *IEEE Transactions on Industrial Electronics* 67 (10 2020), pp. 8723–8731. ISSN: 0278-0046. https://doi.org/10.1109/TIE.2019.2946551.
432. Jia Deng, Wei Dong, Richard Socher, Li-Jia Li, Kai Li, and Li Fei-Fei. "ImageNet: A large-scale hierarchical image database." In: *IEEE Conference on Computer Vision and Pattern Recognition (CVPR)* (Miami, FL, US, 20.06.–25.06.2009). IEEE Conference

on Computer Vision and Pattern Recognition. Piscataway, NJ, US: IEEE, 2009. ISBN: 9781424439928. https://doi.org/10.1109/CVPR.2009.5206848.

433. Li Deng. "The MNIST Database of Handwritten Digit Images for Machine Learning Research [Best of the Web]." In: *IEEE Signal Processing Magazine* 29 (6 2012), pp. 141–142. ISSN: 1053-5888. https://doi.org/10.1109/MSP.2012.2211477.
434. Alex Krizhevsky. "Learning multiple layers of features from tiny images." In: (2009). url: https://www.cs.toronto.edu/~kriz/learning-features-2009-TR.pdf (visited on 01.03.2023).
435. Connor Shorten and Taghi M. Khoshgoftaar. "A survey on Image Data Augmentation for Deep Learning." In: *Journal of Big Data* 6 (1 2019), p. 1106. ISSN: 2196-1115. https://doi.org/10.1186/s40537-019-0197-0.
436. Trishna Raj and David A. Howey. *Path Dependent Battery Degradation Dataset Part 1*. 2020. https://doi.org/10.5287/bodleian:v0ervBv6p. (visited on 01.03.2023).
437. Trishna Raj and David A. Howey. *Path Dependent Battery Degradation Dataset Part 2*. 2021. https://doi.org/10.5287/bodleian:2zvyknyRg. (visited on 01.03.2023).
438. Trishna Raj and David A. Howey. *Path Dependent Battery Degradation Dataset Part 3*. 2021. https://doi.org/10.5287/bodleian:j1a2eD7ow. (visited on 01.03.2023).
439. Claudia Nobis and Tobias Kuhnimhof. *Mobilität in Deutschland—MiD. Ergebnisbericht. BMVI, infas, DLR, IVT, infas 360. Bonn, Berlin*. 2018. url: https://elib.dlr.de/125879/1/MiD2017_Ergebnisbericht.pdf (visited on 01.03.2023).
440. Niklas Hartmann and Enver Doruk Özdemir. "Impact of different utilization scenarios of electric vehicles on the German grid in 2030." In: *Journal of Power Sources* 196 (4 2011), pp. 2311–2318. ISSN: 03787753. https://doi.org/10.1016/j.jpowsour.2010.09.117.
441. Jaap Schijve. "Load Spectra." In: *Fatigue of Structures and Materials*. Ed. by Jaap Schijve. 1st ed. Dordrecht, The Netherlands: Springer Netherlands, 2009, pp. 259–293. ISBN: 9781402068072. https://doi.org/10.1007/978-1-4020-6808-9_9.
442. Shayan Jawed, Josif Grabocka, and Lars Schmidt-Thieme. "Self-supervised Learning for Semi-supervised Time Series Classification." In: *Advances in Knowledge Discovery and Data Mining*. Ed. by Hady W. Lauw, Raymond Chi-Wing Wong, Alexandros Ntoulas, Ee-Peng Lim, See-Kiong Ng, and Sinno Jialin Pan. 12084. Cham, Switzerland: Springer International Publishing, 2020, pp. 499–511. ISBN: 978-3-030-47426-3. https://doi.org/10.1007/978-3-030-47426-3_39.
443. David R. Larson. "Unitary systems and wavelet sets." In: *Wavelet Analysis and Applications*. Ed. by John J. Benedetto, Tao Qian, Mang I. Vai, and Yuesheng Xu. Basel, Switzerland: Birkhäuser Basel, 2007, pp. 143–171. ISBN: 978-3-7643-7778-6. https://doi.org/10.1007/978-3-7643-7778-6_14.
444. Richard Meyes, Johanna Donauer, Andre Schmeing, and Tobias Meisen. "A Recurrent Neural Network Architecture for Failure Prediction in Deep Drawing Sensory Time Series Data." In: *Procedia Manufacturing* 34 (1-2 2019), pp. 789–797. ISSN: 23519789. https://doi.org/10.1016/j.promfg.2019.06.205.
445. Hong-Keun Kim, Charn-Jung Kim, Chang-Wan Kim, and Kyu-Jin Lee. "Numerical analysis of accelerated degradation in large lithium-ion batteries." In: *Computers & Chemical Engineering* 112 (2018), pp. 82–91. ISSN: 1873-4375. https://doi.org/10.1016/j.compchemeng.2017.12.019.

446. Moritz Hardt and Tengyu Ma. "Identity Matters in Deep Learning." In: *International Conference on Learning Representations (ICLR) 2017* (2017). https://doi.org/10.48550/arXiv.1611.04231.
447. Phillip Stoffel, Laura Maier, Alexander Kümpel, Thomas Schreiber, and Dirk Müller. "Evaluation of advanced control strategies for building energy systems." In: *Energy and Buildings* 280 (1 2023), p. 112709. ISSN: 03787788. https://doi.org/10.1016/j.enbuild.2022.112709.
448. Yixin Yang. "A machine-learning prediction method of lithium-ion battery life based on charge process for different applications." In: *Applied Energy* 292 (2021). ISSN: 03062619; 1872-9118. https://doi.org/10.1016/j.apenergy.2021.116897.
449. Nitish Srivastava, Geoffrey E. Hinton, Alex Krizhevsky, Ilya Sutskever, and Ruslan R. Salakhutdinov. "Dropout: A simple way to prevent neural networks from overfitting." In: *Journal of Machine Learning Research* 15 (1 2014), pp. 1929–1958. ISSN: 1532-4435. url: http://jmlr.org/papers/v15/srivastava14a.html (visited on 01.03.2023).
450. Youngwoo Choi, Dohyun Jung, Kyoungchun Ham, and Sungin Bae. "A study on the accelerated vibration endurance tests for battery fixing bracket in electrically driven vehicles." In: *Procedia Engineering* 10 (2011), pp. 851–856. ISSN: 18777058. https://doi.org/10.1016/j.proeng.2011.04.140.
451. Ernst Gassner. "Vereinfachter Betriebsfestigkeits-Nachweis für zufallsartig im hohen Lebensdauerbereich beanspruchte Fahrzeugbauteile / Time-reduced performance fatigue test for automotive components randomly loaded in the high endurance range." In: *Materials Testing* 26 (8 1984), pp. 274–276. ISSN: 2195-8572. https://doi.org/10.1515/mt-1984-260807.
452. Vitor Cerqueira, Luis Torgo, and Igor Mozetič. "Evaluating time series forecasting models. An empirical study on performance estimation methods." In: *Machine Learning* 109 (11 2020), pp. 1997–2028. ISSN: 0885-6125. https://doi.org/10.1007/s10994-020-05910-7.
453. Marcus Johnen, Simon Pitzen, Udo Kamps, Maria Kateri, Philipp André Dechent, and Dirk Uwe Sauer. "Modeling long-term capacity degradation of lithium-ion batteries." In: *Journal of Energy Storage* 34 (9 2021), p. 102011. ISSN: 2352152X. https://doi.org/10.1016/j.est.2020.102011.
454. Claire Sugihara, Katrina Sutton, Adam Davis, Vaishnavi Karanam, and Gil Tal. "From sport to eco. A case study of driver inputs on electric vehicle efficiency." In: *Transportation Research Part F: Traffic Psychology and Behaviour* 82 (5 2021), pp. 412–428. ISSN: 13698478. https://doi.org/10.1016/j.trf.2021.09.007.
455. Jinwen Li, Zhongwei Deng, Hongao Liu, Yi Xie, Chuan Liu, and Chen Lu. "Battery capacity trajectory prediction by capturing the correlation between different vehicles." In: *Energy* 260 (11 2022), p. 125123. ISSN: 1873-6785. https://doi.org/10.1016/j.energy.2022.125123.
456. Peter Keil, Simon F. Schuster, Jörn Wilhelm, Julian Travi, Andreas Hauser, Ralph C. Karl, and Andreas Jossen. "Calendar Aging of Lithium-Ion Batteries." In: *Journal of The Electrochemical Society* 163 (9 2016), A1872-A1880. ISSN: 0013-4651. https://doi.org/10.1149/2.0411609jes.
457. Chrispin Gogoda Mbewe, Richard Blanchard, Gianfranco Claudio, Martin Bliss, Nigel Monk, and John Barton. "Promoting Battery Testing in Developing Countries Through Development of a Low-Cost Battery Test System." In: *IEEE International Humanita-*

rian Technology Conference (IHTC) (Virtual, 02.12.–04.12.2021). Piscataway, NJ, US: IEEE, 2021, pp. 1–4. ISBN: 9781665449267. https://doi.org/10.1109/IHTC53077.2021.9698932.

458. Billy Wu, Widanalage Dhammika Widanage, Shichun Yang, and Xinhua Liu. "Battery digital twins. Perspectives on the fusion of models, data and artificial intelligence for smart battery management systems." In: *Energy and AI* 1 (2020), p. 100016. ISSN: 26665468. https://doi.org/10.1016/j.egyai.2020.100016.
459. Gabriel Michau, Thomas Palmé, and Olga Fink. "Fleet PHM for Critical Systems: Bi-level Deep Learning Approach for Fault Detection." In: *Proceedings of the European Conference of the PHM Society 2018* 4 (1 2018), pp. 403–413. https://doi.org/10.36001/phme.2018.v4i1.403.
460. Robert J. Brown. "A new marketing tool. Life-cycle costing." In: *Industrial Marketing Management* 8 (2 1979), pp. 109–113. ISSN: 00198501. https://doi.org/10.1016/0019-8501(79)90050-6.
461. Alfred Kärcher Vertriebs-GmbH. Kärcher Fleet. 2020. url: https://www.kaercher.com/uk/services/professional/connected-cleaning/kaercher-fleet.html (visited on 01.03.2023).
462. Linde Material Handling GmbH. *The networked fleet. Digital fleet management from Linde*. 2020. url: https://www.linde-mh.com/en/Solutions/Fleet-Management/ (visited on 01.03.2023).
463. MAN Truck & Bus UK Limited. *MAN Fleet Management—Telematics & Workshop Management*. 2020. url: https://www.manfleetmanagement.co.uk/ (visited on 01.03.2023).
464. Volkswagen AG. We *Connect Fleet*. 2020. url: https://www.volkswagen-nutzfahrzeugede/de/services/digitale-services/we-connect-fleet.html (visited on 01.03.2023).
465. Scania CV AB. *Scania Fleet Management. Data drives Development*. 2016. url: https://www.scania.com/content/dam/scanianoe/market/uk/brochures/connectedservices/brochure-scania-fleet-management.pdf (visited on 01.03.2023).
466. Deutsche Bahn AG. *Fuhrparkmanagement*. 2020. url: https://www.deutschebahn.com/de/geschaefte/fahrzeuge/db_connect/Fuhrparkmanagement?contentId=1193366 (visited on 01.03.2023).
467. BayWa AG. *BayWa Mobility Solutions*. 2020. url: https://www.baywa-mobility.de/de/emobility/produkte-und-services/flottenanalyse (visited on 01.03.2023).
468. Webfleet Solutions B.V. *Fleet Management*. 2020. url: https://www.webfleet.com/en_gb/webfleet/fleet-management/ (visited on 01.03.2023).
469. Shawn Midlam-Mohler, Sean Ewing, Vincenzo Marano, Yann Guezennec, and Giorgio Rizzoni. "PHEV fleet data collection and analysis." In: *IEEE Vehicle Power and Propulsion Conference (VPPC)* (Dearborn, MI, US, 07.09.–10.09.2009). 2009, pp. 1205–1210. ISBN: 9781424426003. https://doi.org/10.1109/VPPC.2009.5289712.
470. Pierre Lenain, M. Kechmire, and J. P. Smaha. "Monitoring fleets of electric vehicles. Optimizing operational use and maintenance." In: *Journal of Power Sources* 53 (2 1995), pp. 335–338. ISSN: 03787753. https://doi.org/10.1016/0378-7753(94)01995-8.
471. Markus Weinberger, Felix Wortmann, Dominik Bilgeri, and Elgar Fleisch. "Bosch-Flottenmanagement: Das IoT fordert die Organisation." In: *Digitale Transformation gestalten. Geschäftsmodelle, Erfolgsfaktoren, Checklisten*. Ed. by Oliver Gassmann and

Philipp Sutter. 2nd ed. 2019, pp. 217–224. ISBN: 978-3-446-45868-0. https://doi.org/10.3139/9783446459632.016.

472. Gernot Lobenberg and Stefan Friedel. "Flottenmanagement im Straßengüterverkehr." In: *Kompendium der Verkehrstelematik. Technologien, Applikationen, Perspektiven*. Ed. by Harry Evers. 4th ed. Köln, Germany: Verl. TÜV Rheinland, 1998, p. 7410. ISBN: 3-8249-0555-8.
473. Asvin Goel and Volker Gruhn. "A Fleet Monitoring System for Advanced Tracking of Commercial Vehicles." In: *IEEE International Conference on Systems, Man, and Cybernetics (SMC)* (Taipei, Taiwan, 08.10.–11.10.2006). Piscataway, NJ, US: IEEE Service Center, 2007, pp. 2517–2522. ISBN: 9781424400997. https://doi.org/10.1109/ICSMC.2006.385242.
474. Claus Grøn Sørensen and D. D. Bochtis. "Conceptual model of fleet management in agriculture." In: *Biosystems Engineering* 105 (1 2010), pp. 41–50. ISSN: 15375110. https://doi.org/10.1016/j.biosystemseng.2009.09.009.
475. Gangyan Xu, Ming Li, Lizi Luo, Chun-Hsien Chen, and George Q. Huang. "Cloud-based fleet management for prefabrication transportation." In: *Enterprise Information Systems* 13 (1 2018), pp. 87–106. ISSN: 1751-7575. https://doi.org/10.1080/17517575.2018.1455109.
476. Sean Bennett. *Heavy duty truck systems*. 7th ed. Boston, MA, US: Cengage, 2020. 1 online resource. ISBN: 1337787108.
477. Holger Billhardt, Alberto Fernandez, Lissette Lemus, Marin Lujak, Nardine Osman, Sascha Ossowski, and Carles Sierra. "Dynamic Coordination in Fleet Management Systems. Toward Smart Cyber Fleets." In: *IEEE Intelligent Systems* 29 (3 2014), pp. 70–76. ISSN: 1541-1672. https://doi.org/10.1109/MIS.2014.41.
478. Michal Levi-Bliech, Polina Kurtser, Nava Pliskin, and Lior Fink. "The effects of a fleet-Management App on Driver Behavior." In: *26th European Conference on Information Systems (ECIS2018)* (Portsmouth, UK, 26.06.–28.06.2018). 2018.
479. Andrea Attanasio, Jay Bregman, Gianpaolo Ghiani, and Emanuele Manni. "Real-Time Fleet Management At Ecourier Ltd." In: *Dynamic Fleet Management* . Ed. by Ramesh Sharda, Stefan Voß, Vasileios Zeimpekis, Christos D. Tarantilis, George M. Giaglis, and Ioannis Minis. 1st ed. 38. Boston, MA, US: Springer US, 2007, pp. 219–238. ISBN: 978-0-387-71721-0.
480. Jim Mele. *Fleet Management Systems: The future is here*. 2005. url: https://www.fleetowner.com/technology/article/21662608/fleet-management-systems-the-future-is-here (visited on 01.03.2023).
481. Branimir Škugor and Joško Deur. "A bi-level optimisation framework for electric vehicle fleet charging management." In: *Applied Energy* 184 (2016), pp. 1332–1342. ISSN: 03062619; 1872-9118. https://doi.org/10.1016/j.apenergy.2016.03.091.
482. Junjie Hu, Hugo Morais, Tiago Sousa, and Morten Lind. "Electric vehicle fleet management in smart grids. A review of services, optimization and control aspects." In: *Renewable and Sustainable Energy Reviews* 56 (2016), pp. 1207–1226. ISSN: 13640321. https://doi.org/10.1016/j.rser.2015.12.014.
483. Maxime Monnin, Bouthaina Abichou, Alexandre Voisin, and Christophe Mozzati. "Fleet historical cases for predictive maintenance." In: *International Conference on acoustical and vibratory methods in surveillance and diagnostics, Surveillan-*

ce 6 (Margny-lès-Compiègne, France, 25.10.–26.10.2011). 2011. url: https://hal.archivesouvertes.fr/hal-00638399.

484. Renato Giorgiani do Nascimento and Felipe A. C. Viana. "Fleet Prognosis with Physics-informed Recurrent Neural Networks." In: *The 12th International Workshop on Structural Health Monitoring* (Stanford, CA, US, 10.09.–12.09.2019). 2019. https://doi.org/10.12783/shm2019/32301.
485. Alexandre Voisin, Gabriela Medina-Oliva, Maxime Monnin, Jean-Baptiste Léger, and Benoît Iung. "Fleet-wide Diagnostic and Prognostic Assessment." In: *Annual Conference of the Prognostics and Health Management Society* 5 (1 2013). https://doi.org/10.36001/phmconf.2013.v5i1.2311.
486. Deutsches Institut für Normung e.V. DIN EN 62264-1:2014-07. 2014. *Integration von Unternehmensführungs- und Leitsystemen- Teil 1: Modelle und Terminologie.*
487. Timothy Kaufmann and Hans-Gerd Servatius. "IoT- und KI-Architektur." In: *Das Internet der Dinge und Künstliche Intelligenz als Game Changer*. Ed. by Timothy Kaufmann and Hans-Gerd Servatius. 1st ed. Wiesbaden, Germany: Springer Fachmedien Wiesbaden, 2020, pp. 113–149. ISBN: 978-3-658-28399-5. https://doi.org/10.1007/978-3-658-28400-8_8.
488. Marcus Adams, Wolfang Kühn, Thomas Stör, and Martin Zelm. "DIN EN 62264. Die neue Norm zur Interoperabilität von Produktion und Unternehmensführung—Teil 1." In: *Automatisierungstechnische Praxis* 49 (5 2007), pp. 52–57.
489. Edward Glaessgen and David Stargel. "The Digital Twin Paradigm for Future NASA and U.S. Air Force Vehicles." In: *53rd Structures, Structural Dynamics, and Materials Conference: Special Session on the Digital Twin* (Honolulu, HI, US, 23.04.–26.04.2012). 2012. https://doi.org/10.2514/6.2012-1818.
490. Massimo Merenda, Carlo Porcaro, and Demetrio Iero. "Edge Machine Learning for AI-Enabled IoT Devices: A Review." In: *Sensors* 20 (9 2020), pp. 1–34. https://doi.org/10.3390/s20092533.
491. Alberto Savoia. *Right It. How to Beat the Law of Market Failure*. 1st ed. New York, NY, US: HarperCollins Publishers, 2019. 272 pp. ISBN: 978-0062884657.
492. Suhrid Satyal, Ingo Weber, Hye-Young Paik, Claudio Di Ciccio, and Jan Mendling. "Shadow Testing for Business Process Improvement." In: *On the Move to Meaningful Internet Systems*. Ed. by Hervé Panetto, Christophe Debruyne, Henderik A. Proper, Claudio Agostino Ardagna, Dumitru Roman, and Robert Meersman. 1st ed. Programming and Software Engineering 11229. Cham, Switzerland: Springer, 2018, pp. 153–171. ISBN: 978-3-030-02610-3. https://doi.org/10.1007/978-3-030-02610-3_9.
493. Wolfgang Bernhart. "Wertschöpfung steigern mit Battery-as-a-Service." In: *ATZextra* 25 (S5 2020), pp. 14–19. ISSN: 2195-1454. https://doi.org/10.1007/s35778-020-0141-0.
494. Qiang Yang, Yang Liu, Tianjian Chen, and Yongxin Tong. "Federated Machine Learning." In: *ACM Transactions on Intelligent Systems and Technology* 10 (2 2019), pp. 1–19. ISSN: 2157-6904. https://doi.org/10.1145/3298981.
495. Peter Kairouz, H. Brendan McMahan, Brendan Avent, Aurélien Bellet, Mehdi Bennis, Arjun Nitin Bhagoji, Kallista Bonawitz, Zachary Charles, Graham Cormode, Rachel Cummings, Rafael G. L. D'Oliveira, Hubert Eichner, Salim El Rouayheb, David Evans, Josh Gardner, Zachary Garrett, Adrià Gascòn, Badih Ghazi, Phillip B. Gibbons, Marco Gruteser, Zaid Harchaoui, Chaoyang He, Lie He, Zhouyuan Huo, Ben Hutchinson,

Justin Hsu, Martin Jaggi, Tara Javidi, Gauri Joshi, Mikhail Khodak, Jakub Konečný, Aleksandra Korolova, Farinaz Koushanfar, Sanmi Koyejo, Tancrède Lepoint, Yang Liu, Prateek Mittal, Mehryar Mohri, Richard Nock, Ayfer Özgür, Rasmus Pagh, Hang Qi, Daniel Ramage, Ramesh Raskar, Mariana Raykova, Dawn Song, Weikang Song, Sebastian U. Stich, Ziteng Sun, Ananda Theertha Suresh, Florian Tramèr, Praneeth Vepakomma, Jianyu Wang, Li Xiong, Zheng Xu, Qiang Yang, Felix X. Yu, Han Yu, and Sen Zhao. "Advances and Open Problems in Federated Learning." In: *Foundations and Trends® in Machine Learning* 14 (1–2 2021), pp. 1–210. ISSN: 1935-8237. https://doi.org/10.1561/2200000083.

496. Jakub Konečný, H. Brendan McMahan, and Daniel Ramage. *Federated Optimization: Distributed Optimization Beyond the Datacenter*. 2015. https://doi.org/10.48550/arXiv.1511.03575.
497. Jakub Konečný, H. Brendan McMahan, Felix X. Yu, Peter Richtárik, Ananda Theertha Suresh, and Dave Bacon. *Federated Learning: Strategies for Improving Communication Efficiency*. NIPSWorkshop on Private Multi-Party Machine Learning, 2017. https://doi.org/10.48550/arXiv.1610.05492.
498. George Baure, Arnaud Devie, and Matthieu Dubarry. "Battery Durability and Reliability under Electric Utility Grid Operations. Path Dependence of Battery Degradation." In: *Journal ofThe Electrochemical Society* 166 (10 2019), A1991-A2001. ISSN: 0013-4651. https://doi.org/10.1149/2.0971910jes.
499. Dzmitry Bahdanau, Kyunghyun Cho, and Yoshua Bengio. "Neural Machine Translation by Jointly Learning to Align and Translate." In: *Proceedings of the 3rd International Conference on Learning Representations (ICLR)* (2015). https://doi.org/10.48550/arXiv.1409.0473.
500. Felix Heinrich, Fabian Kai-Dietrich Noering, Marco Pruckner, and Konstantin Jonas. "Unsupervised data-preprocessing for Long Short-Term Memory based battery model under electric vehicle operation." In: *Journal of Energy Storage* 38 (5 2021), p. 102598. ISSN: 2352152X. https://doi.org/10.1016/j.est.2021.102598.
501. Arka Daw, Anuj Karpatne, William D. Watkins, Jordan S. Read, and Vipin Kumar. "Physics-Guided Neural Networks (PGNN). An Application in Lake Temperature Modeling." In: *Knowledge-Guided Machine Learning*. Ed. by Anuj Karpatne, Ramakrishnan Kannan, and Vipin Kumar. Boca Raton, FL, US: Chapman and Hall/CRC, 2022, pp. 353–372. ISBN: 9781003143376. https://doi.org/10.1201/9781003143376-15.
502. Artur M. Schweidtmann and Alexander Mitsos. "Deterministic Global Optimization with Artificial Neural Networks Embedded." In: *Journal of Optimization Theory and Applications* 180 (3 2019), pp. 925–948. ISSN: 0022-3239. https://doi.org/10.1007/s10957-018-1396-0.
503. Stephen Boyd and Lieven Vandenberghe. *Convex optimization. The science of the small in physics, engineering, chemistry, biology and medicine*. Vol. 13:1-2. Foundations and trends in theoretical computer science. New Delhi, India: Cambridge University Press, 2004. xiii, 716. ISBN: 9781316603598.
504. John Cannarella and Craig B. Arnold. "Stress evolution and capacity fade in constrained lithium-ion pouch cells." In: *Journal of Power Sources* 245 (2014), pp. 745–751. ISSN: 03787753. https://doi.org/10.1016/j.jpowsour.2013.06.165.
505. Wenhua Li, Zhipeng Jiao, Qian Xiao, Jinhao Meng, Yunfei Mu, Hongjie Jia, Remus Teodorescu, and Frede Blaabjerg. "A Study on Performance Characterization Consi-

dering Six-Degree-of-Freedom Vibration Stress and Aging Stress for Electric Vehicle Battery Under Driving Conditions." In: *IEEE Access* 7 (2019), pp. 112180–112190. ISSN: 2169-3536. https://doi.org/10.1109/ACCESS.2019.2935380.

506. Severin Hahn, Sebastian Theil, Jan Kroggel, and Kai Peter Birke. "Pressure Prediction Modeling and Validation for Lithium-Ion Pouch Cells in Buffered Module Assemblies." In: *Journal of Energy Storage* 40 (2021). ISSN: 2352152X. https://doi.org/10.1016/j.est.2021.102517.
507. Xia Hua and Alan Thomas. "Effect of dynamic loads and vibrations on lithium-ion batteries." In: *Journal of Low Frequency Noise, Vibration and Active Control* 40 (4 2021), pp. 1927–1934. ISSN: 1461-3484. https://doi.org/10.1177/14613484211008112.
508. Karl Lowenmark, Cees Taal, Joakim Nivre, Marcus Liwicki, and Fredrik Sandin. "Processing of Condition Monitoring Annotations with BERT and Technical Language Substitution. A Case Study." In: *PHM Society European Conference* 7 (1 2022), pp. 306–314. https://doi.org/10.36001/phme.2022.v7i1.3356.
509. Council of the European Union. *Proposal for a Regulation of the European Parliament and of the Council concerning batteries and waste batteries, repealing Directive 2006/66/EC and amending Regulation (EU)* No 2019/1020. 2022. url: https://data.consilium.europa.eu/doc/document/ST-7103-2022-REV-1/en/pdf (visited on 01.03.2023).
510. James T. Frith, Matthew J. Lacey, and Ulderico Ulissi. "A non-academic perspective on the future of lithium-based batteries." In: *Nature Communications* 14 (1 2023), p. 420. ISSN: 2041-1723. https://doi.org/10.1038/s41467-023-35933-2.
511. Xuansen Fang, Yaolong He, Xiaomin Fan, Dan Zhang, and Hongjiu Hu. "Modeling and Simulation in Capacity Degradation and Control of All-Solid-State Lithium Battery Based on Time-Aging Polymer Electrolyte." In: *Polymers* 13 (8 2021). https://doi.org/10.3390/polym13081206.
512. Nora Manthey. *All Solid-State Battery: Could Nissan make it first?* 2022. url: https://www.electrive.com/2022/04/18/nissan-all-solid-state-battery-will-nissanmake-it-first/ (visited on 01.03.2023).
513. International Energy Agency. *Global Electric Vehicle Outlook 2022*. 2022. url: https://iea.blob.core.windows.net/assets/ad8fb04c-4f75-42fc-973a-6e54c8a4449a/GlobalElectricVehicleOutlook2022.pdf (visited on 01.03.2023).
514. Nationale Leitstelle Ladeinfrastruktur. Einfach laden an Rastanlagen. Auslegung des Netzanschlusses für E-Lkw-Lade-Hubs. Ed. by Nationale Leitstelle Ladeinfrastruktur. 2022. url: https://nationale-leitstelle.de/wp-content/uploads/2022/09/LeitstellespsLKW-Netzstudie.pdf (visited on 01.03.2023).
515. ABB Switzerland Ltd. *ABB Review 1/2023*. 2023. url: https://search.abb.com/library/Download.aspx?DocumentID=9AKK108467A6063&LanguageCode=en&DocumentPartId=&Action=Launch (visited on 01.03.2023).
516. Pyeong-Yeon Lee, Sanguk Kwon, Deokhun Kang, Inho Cho, and Jonghoon Kim. "Principle component analysis-based optimized feature extraction merged with nonlinear regression model for improved state-of-health prediction." In: *Journal of Energy Storage* 48 (10 2022), p. 104026. ISSN: 2352152X. https://doi.org/10.1016/j.est.2022.104026.
517. Nipat Rasmekomen and Ajith K. Parlikad. "Optimising Maintenance of Multi- Component Systems with Degradation Interactions." In: *IFAC Proceedings Volumes* 47 (3

2014), pp. 7098–7103. ISSN: 14746670. https://doi.org/10.3182/20140824-6-ZA-1003.01447.

518. Phuc Do, Roy Assaf, Phil Scarf, and Benoît Iung. "Modelling and application of condition-based maintenance for a two-component system with stochastic and economic dependencies." In: *Reliability Engineering and System Safety* 182 (28 2019), pp. 86–97. ISSN: 1879-0836. https://doi.org/10.1016/j.ress.2018.10.007.
519. Van-Thai Nguyen, Phuc Do, and Alexandre Voisin. "Artificial-Intelligence-Based Maintenance Scheduling for Complex Systems with Multiple Dependencies." In: *PHM Society European Conference* 7 (1 2022), pp. 586–589. https://doi.org/10.36001/phme.2022.v7i1.3294.
520. Huaibin Wang, Shuyu Wang, Xuning Feng, Xuan Zhang, Kangwei Dai, Jun Sheng, Zhenyang Zhao, Zhiming Du, Zelin Zhang, Kai Shen, Chengshan Xu, QinzhengWang, Xiaoyu Sun, Yanliang Li, Jiaju Ling, Jiani Feng, Hewu Wang, and Minggao Ouyang. "An experimental study on the thermal characteristics of the Cell-To-Pack system." In: *Energy* 227 (2021), p. 120338. ISSN: 1873-6785. https://doi.org/10.1016/j.energy.2021.120338.
521. Eduard Gerlitz, David Botzem, Hannes Weinmann, Janna Ruhland, and Jürgen Fleischer. "Cell-to-Pack-Technologie für Li-Ionen-Batterien." In: *Zeitschrift für wirtschaftlichen Fabrikbetrieb* 116 (10 2021), pp. 689–694. ISSN: 2511-0896. https://doi.org/10.1515/zwf-2021-0146.
522. Leo Wildfeuer, Michael Baumann, and Markus Lienkamp. "Concept for a hybrid semiempirical and data-based model for state-of-health monitoring and aging prediction of li-ion battery packs." In: *Advanced Automotive Batteries Conference Europe (AABC)* (Strasbourg, France, 24.06.–27.06.2019). 2019. https://doi.org/10.13140/rg.2.2.29057.51043.
523. German I. Parisi, Ronald Kemker, Jose L. Part, Christopher Kanan, and Stefan Wermter. "Continual lifelong learning with neural networks: A review." In: *Neural Networks* 113 (2019), pp. 54–71. ISSN: 08936080. https://doi.org/10.1016/j.neunet.2019.01.012.
524. Xing Shu, Shiquan Shen, Jiangwei Shen, Yuanjian Zhang, Guang Li, Zheng Chen, and Yonggang Liu. "State of health prediction of lithium-ion batteries based on machine learning. Advances and perspectives." In: *iScience* 24 (11 2021), p. 103265. https://doi.org/10.1016/J.ISCI.2021.103265.
525. Yoshua Bengio, Jérôme Louradour, Ronan Collobert, and Jason Weston. "Curriculum learning." In: *Proceedings of the 26th Annual International Conference on Machine Learning*. International Conference on Machine Learning (ICML) (Montreal, Quebec, Canada, 14.06.–18.06.2009). Ed. by Andrea Danyluk. New York, NY, US: Association for Computing Machinery Inc. (ACM), 2009, pp. 41–48. ISBN: 978-1-60558-516-1. https://doi.org/10.1145/1553374.1553380.
526. Pang Wei Koh and Percy Liang. "Understanding Black-box Predictions via Influence Functions." In: *34th International Conference on Machine Learning (ICML 2017)*. International Conference on Machine Learning (ICML) (Sydney, Australia, 06.08.–11.08.2017). Ed. by Doina Precup and Yee Whye Teh. 70. Stroudsburg, PA, US: International Machine Learning Society (IMLS), 2017, pp. 1885–1894. ISBN: 9781510855144. url: http://proceedings.mlr.press/v70/koh17a/koh17a.pdf.
527. Kai Vuorilehto. "Materials and function." In: *Lithium-Ion Batteries: Basics and Applications*. Ed. by Reiner Korthauer. Berlin, Heidelberg, Germany: Springer Berlin Heidel-

berg, 2018, pp. 21–28. ISBN: 978-3-662-53069-6. https://doi.org/10.1007/978-3-662-53071-9_3.

528. Christian Graf. "Cathode materials for lithium-ion batteries." In: *Lithium-Ion Batteries: Basics and Applications*. Ed. by Reiner Korthauer. Berlin, Heidelberg, Germany: Springer Berlin Heidelberg, 2018, pp. 29–41. ISBN: 978-3-662-53069-6. https://doi.org/10.1007/978-3-662-53071-9_4.
529. Jennifer Schoch. "Battery Life Optimal Operation of Electric Vehicles." PhD Thesis. Karlsruhe, Germany, 2018. 130 pp. url: https://publikationen.bibliothek.kit.edu/1000083119/13974324.
530. Christoph Hartnig and Michael Schmidt. "Electrolytes and conducting salts." In: *Lithium-Ion Batteries: Basics and Applications*. Ed. by Reiner Korthauer. Berlin, Heidelberg, Germany: Springer Berlin Heidelberg, 2018, pp. 59–74. ISBN: 978-3-662-53069-6.
531. Christoph J.Weber and Michael Roth. "Separators." In: *Lithium-Ion Batteries: Basics and Applications*. Ed. by Reiner Korthauer. Berlin, Heidelberg, Germany: Springer Berlin Heidelberg, 2018, pp. 75–88. ISBN: 978-3-662-53069-6. https://doi.org/10.1007/978-3-662-53071-9_7.
532. Thomas Woehrle. "Lithium-ion cell." In: *Lithium-Ion Batteries: Basics and Applications*. Ed. by Reiner Korthauer. Berlin, Heidelberg, Germany: Springer Berlin Heidelberg, 2018, pp. 101–111. ISBN: 978-3-662-53069-6. https://doi.org/10.1007/978-3-662-53071-9_9.
533. International Organization for Standardization. ISO/IEC 61960-3:2017. 2017. *Secondary cells and batteries containing alkaline or other non-acid electrolytes – Secondary lithium cells and batteries for portable applications – Part 3: Prismatic and cylindrical lithium secondary cells and batteries made from them.*
534. Lydia Dorrmann, Kerstin Sann-Ferro, Patrick Heininger, and Jochen Mähli. *Kompendium: Li-Ionen-Batterien. Grundlagen, Merkmale, Gesetze und Normen.* Ed. by Verband der Elektrotechnik Elektronik Informationstechnik e. V. 2021. url: https://www.dke.de/resource/blob/933404/fa7a24099c84ef613d8e7afd2c860a39/kompendium-li-ionen-batterien-data.pdf (visited on 01.03.2023).
535. Hong-Keun Kim, Jin Hyeok Choi, and Kyu-Jin Lee. "A Numerical Study of the Effects of Cell Formats on the Cycle Life of Lithium Ion Batteries." In: *Journal of The Electrochemical Society* 166 (10 2019), A1769-A1778. ISSN: 0013-4651. https://doi.org/10.1149/2.0261910jes.
536. Xiaosong Hu, Hao Yuan, Changfu Zou, Zhe Li, and Lei Zhang. "Co-Estimation of State of Charge and State of Health for Lithium-Ion Batteries Based on Fractional-Order Calculus." In: *IEEE Transactions on Vehicular Technology* 67 (11 2018), pp. 10319–10329. ISSN: 0018-9545. https://doi.org/10.1109/TVT.2018.2865664.
537. M. Nisvo Ramadan, Bhisma Adji Pramana, Sigit Agung Widayat, Lora Khaula Amifia, Adha Cahyadi, and Oyas Wahyunggoro. "Comparative Study Between Internal Ohmic Resistance and Capacity for Battery State of Health Estimation." In: *Journal of Mechatronics, Electrical Power, and Vehicular Technology* 6 (2 2015), pp. 113–122. ISSN: 2088-6985. https://doi.org/10.14203/j.mev.2015.v6.113-122.
538. Jürgen Remmlinger, Michael Buchholz, Thomas Soczka-Guth, and Klaus Dietmayer. "On-board state-of-health monitoring of lithium-ion batteries using linear parameterva-

rying models." In: *Journal of Power Sources* 239 (9 2013), pp. 689–695. ISSN: 03787753. https://doi.org/10.1016/j.jpowsour.2012.11.102.

539. Volkswagen AG. *Key components for a new era—the battery system*. 2019. url: https://www.volkswagenag.com/en/news/2019/12/key-components-for-a-new-era--the-battery-system.html (visited on 01.03.2023).
540. Gareth James, Daniela Witten, Trevor Hastie, and Robert Tibshirani. *An Introduction to Statistical Learning*. Vol. 103. New York, NY, US: Springer, 2013. ISBN: 978-1-4614-7137-0. https://doi.org/10.1007/978-1-4614-7138-7.
541. Stephen Wu, Sijia Liu, Sunghwan Sohn, Sungrim Moon, Chung-Il Wi, Young Juhn, and Hongfang Liu. "Modeling asynchronous event sequences with RNNs." In: *Journal of Biomedical Informatics* 83 (2018), pp. 167–177. ISSN: 1532-0480. https://doi.org/10.1016/j.jbi.2018.05.016.
542. Antti Aitio and David A. Howey. "Combining Non-Parametric and Parametric Models for Stable and Computationally Efficient Battery Health Estimation." In: *Proceedings of the ASME 2020 Dynamic Systems and Control Conference (DSCC2020), Volume 1* (Online, 05.10.–07.10.2020). New York, NY, US: The American Society of Mechanical Engineers, 2020. ISBN: 978-0-7918-8427-0. https://doi.org/10.1115/DSCC2020-3180.
543. Alex Graves. "Practical Variational Inference for Neural Networks." In: *25th Annual Conference on Neural Information Processing Systems 2011* (Granada, Spain, 12.12.–15.12.2011). Ed. by J. Shawe-Taylor, R. Zemel, P. Bartlett, F. Pereira, and Kilian Q. Weinberger. 24. Curran Associates Inc., 2011. ISBN: 978-1-61839-599-3. url: http://papers.nips.cc/paper/4329-practical-variational-inference-for-neural-networks.pdf.
544. Daler Rakhmatov, Sarma Vrudhula, and Deborah A. Wallach. "A model for battery lifetime analysis for organizing applications on a pocket computer." In: *IEEE Transactions on Very Large Scale Integration (VLSI) Systems* 11 (6 2003), pp. 1019–1030. ISSN: 1063-8210. https://doi.org/10.1109/TVLSI.2003.819320.
545. Lars Schnieder. *Strategisches Management von Fahrzeugflotten im öffentlichen Personenverkehr.* 1st ed. Berlin, Germany: Springer, 2018. 263 pp. https://doi.org/10.1007/978-3-662-56608-4.
546. Chao Jin, Dragan Djurdjanovic, Hossein D. Ardakani, Keren Wang, Matthew Buzza, Behrad Begheri, Patrick Brown, and Jay Lee. "A comprehensive framework of factory-to-factory dynamic fleet-level prognostics and operation management for geographically distributed assets." In: *IEEE International Conference on Automation Science and Engineering (CASE)* (Gothenburg, Sweden, 24.08.–28.08.2015). 2015, pp. 225–230. https://doi.org/10.1109/CoASE.2015.7294066.
547. Sini-Kaisu Kinnunen, Salla Marttonen-Arola, Antti Ylä-Kujala, Timo Kärri, Toni Ahonen, Pasi Valkokari, and David Baglee. "Decision Making Situations Define Data Requirements in Fleet Asset Management." In: *Proceedings of the 10th World Congress on Engineering Asset Management (WCEAM 2015)*. Ed. by Kari T. Koskinen, Helena Kortelainen, Jussi Aaltonen, Teuvo Uusitalo, Kari Komonen, Joseph Mathew, and Jouko Laitinen. 1st ed. Lecture notes in mechanical engineering. Cham, Switzerland: Springer, 2016, pp. 357–364. ISBN: 978-3-319-27064-7. https://doi.org/10.1007/978-3-319-27064-7_33.
548. Timothy Verstraeten and Ann Nowé. "Reinforcement Learning for Fleet Applications using Coregionalized Gaussian Processes." In: *The Adaptive Learning Agents Workshop at AAMAS* (Stockholm, Sweden, 14.07.–15.07.2018). 2018.

549. Giacomo Leone, Loredana Cristaldi, and Simone Turrin. "A data-driven prognostic approach based on statistical similarity. An application to industrial circuit breakers." In: *Meausrement* 108 (2017), pp. 163–170. ISSN: 02632241. https://doi.org/10.1016/j.measurement.2017.02.017.
550. Karl Weiss, Taghi M. Khoshgoftaar, and DingDing Wang. "A survey of transfer learning." In: *Journal of Big Data* 3 (1 2016), p. 1817. ISSN: 2196-1115. https://doi.org/10.1186/s40537-016-0043-6.
551. Kilian Hendrickx, Wannes Meert, Yves Mollet, Johan Gyselinck, Bram Cornelis, Konstantinos Gryllias, and Jesse Davis. "A general anomaly detection framework for fleet-based condition monitoring of machines." In: *Mechanical Systems and Signal Processing* 139 (2020), p. 106585. ISSN: 08883270. https://doi.org/10.1016/j.ymssp.2019.106585.
552. Volkswagen AG. *ID.3*. 2021. url: https://www.volkswagen.co.uk/electric/electriccars/id3.html (visited on 01.03.2023).

The manufacturer's authorised representative in the EU is Springer Nature Customer Service Centre GmbH, Europaplatz 3, 69115 Heidelberg, Germany. If you have any concerns regarding our products, please contact ProductSafety@springernature.com

Printed and bound by CPI Group (UK) Ltd, Croydon, CR0 4YY
30/09/2026
02225301-0001